# Biomass

edited by
**Maggie Momba and Faizal Bux**

**Biomass**

Edited by Maggie Momba and Faizal Bux

This is a reprint of the book published in 2010 by Sciyo

**Republished by InTech**

Janeza Trdine 9, 51000 Rijeka, Croatia



**Publishing Process Manager** Ana Nikolic

**Technical Editor** Teodora Smiljanic

**Cover Designer** Martina Sirotic

Additional hard copies can be obtained from orders@intechopen.com

Biomass, Edited by Maggie Momba and Faizal Bux
p. cm.
ISBN 978-953-307-113-8

For EU product safety concerns:
IN TECH d.o.o., Prolaz Marije Krucifikse Kozulić 3, 51000 Rijeka, Croatia, info@intechopen.com or visit our website at intechopen.com.

# Contents

# Preface

Due to demands placed on natural resources globally and subsequent deterioration of the environment, there is a need to source and develop appropriate technology to satisfy this requirement. For decades mankind has largely depended on natural resources such fossil fuels to meet the ever increasing energy demands. Realising the finite nature of these resources, emphasis is now shifting to investigating alternate energy source governed by environmentally friendly principles. The abundance of biomass and associated favourable techno-economics has recently changed global perceptions of harnessing biomass as a valuable resource rather than a waste. To this end this book aims to make a contribution to further exploring this area of biomass research and development in the form of a compilation of articles and covering areas of ecological status of different types of biomass and the roles they play in ecosystems, current status of biomass utilization and deriving energy and other value added products from biomass. In this context biomass can be defined as large plants and trees and different groups of microorganisms.

Understanding the ecology of plants and trees especially in forest is imperative for sustainable development. Accurately quantifying forest biomass is of crucial importance for climate change studies particularly with reference to carbon sequestration. Therefore this book highlights the reliable estimation of biomass in forest communities which remains a key challenge for scientists and environmental managers for the successful implementation of sustainable management plan. Diminishing oil and coal reserves has stimulated scientists to focus on biomass such as wood particularly logging residues as a possible source of alternate energy. In addition, the utilisation of renewable lignocellulosic biomass resources will play an increasing role in the future whereby fast pyrolysis of biomass can result in the production of high grade liquid fuels or special chemicals.

The microbial world has also a major role to play as a potential source of biomass. Microorganisms such as microalgae have recently drawn much attention as a potential fuel source and positively impact on carbon dioxide sequestration especially from flue gas, which contribute largely to greenhouse emissions. Although still at developmental stage, microbial fuel cell (MFC) technology has phenomenal potential in the conversion of substrates into electrical energy by microbial catalytic reactions. Organisms that are commonly used in MFC technology include *Geobacter* sp. and *Clostridium* sp. Protozoa have long been known to play a predatory role in conventional wastewater treatment processes. More recently research has shown that they also contribute in the removal of nutrients and pathogenic bacteria from wastewater streams.

This book will serve as an invaluable resource for scientists and environmental managers in planning solutions for sustainable development.

Editors

**Maggie Momba and Faizal Bux**
*Department of Environmental, Water and Earth Sciences,*
*South Africa*

1

# Cordgrass Biomass in Coastal Marshes

Jesús M. Castillo, Alfredo E. Rubio-Casal and Enrique Figueroa
*Universidad de Sevilla*
*Spain*

## 1. Introduction

Cordgrasses (Genus *Spartina*) are one of the most abundant and geographically wide spread halophytes, due to natural and human-mediated dispersal. *S. alterniflora* Loisel, *S. patens* (Aiton) Muhl., *S. spartinae* (Trin.) Merr. ex Hitchc. and *S. cynosuroides* (L.) Roth are native from coastal salt marshes along the East Coast of the American continent, *S. foliosa* Trin. is native to South Californian salt marshes, *S. densiflora* Brongn. and *S. argentinensis* Parodi grow naturally in South American salt marshes and *S. maritima* (Curtis) Fernald, *S. versicolor* Fabre, *S. x townsendii* Groves and *S. anglica* C.E. Hubbard are autochthonous species from European estuaries (Fabre, 1849; Moberley, 1956). Cordgrasses have been introduced to distant salt marshes where they usually behave as invaders. For example, *S. densiflora* colonizes as an alien species the West Coast of North America from San Francisco Bay to British Columbia, the West Coast of Morocco and the Gulf of Cadiz at Southwest Iberian Peninsula (Bortolus, 2006).

Cordgrasses are able to colonize contrasted environments throughout the intertidal gradient such as low marshes and salt pans, and along river channels such as sand spits at river mouths and brackish wetlands landwards. These different coastal marsh habitats show high abiotic and biotic environmental heterogeneity and cordgrasses exhibit a high level of phenotypic plasticity (e.g. Thompson, 1991; Thompson et al., 1991; Trnka & Zedler, 2000; Castillo et al., 2005a). Moreover, some *Spartina* species have developed ecotypes during the process of adaptation to different environmental conditions (e.g. Seliskar et al., 2002; Álvarez et al., 2010). It is also frequent that invasive and native *Spartina* species hybridized. Hybrids may become new allopolyploid species such as *S. anglica* (Schierenbeck & Ellstrand, 2009) and they can develop transgressive traits (e.g. Castillo et al., 2010).

Salt marshes are among the most productive ecosystems in spite of a low plant species richness due to a very stressful environment related with long flooding periods and high salinities (Adam, 1990). Living in this extreme environment, cordgrasses are able to develop dense tussocks, clumps and prairies that may accumulate very high below- and above-ground biomass.

The biomass of cordgrasses plays a very important role in the functioning of salt marshes and estuaries. For example, it controls the development of ecological succession, organizes space occupation of other plants and animals, and plays a key role in estuarine food webs. This chapter analyses inter- and intra-specific variations in the biomass of cordgrasses growing in coastal marshes all around the world.

Fig. 1. Small cordgrass (*Spartina maritima*) grows on intertidal mudflats in European salt marshes where it forms continuous prairies with its coalescent clumps that colonize bare sediments with long rhizomes.

## 2. Sampling cordgrass biomass

Cordgrasses colonize surrounding sediments by successive series of rhizomes, which is reflected in concentric and alternating rings of live and dead shoots with low and high densities and usually in central die-back areas that may remain occupied by necromass. This clonal growth has been described for cordgrasses such as *S. maritima* (Caldwel, 1957; Castellanos et al., 1994) and *S. densiflora* (Castillo et al., 2003). Sediments without vegetation or colonized by other plants are found between expanding *Spartina* clumps prior to their coalescence.

Abiotic factors such as topography, oxygen concentration in sediments, drainage, photoperiod or salinity determine a high degree of environmental heterogeneity at large, medium and small scales (microhabitats) in coastal marshes, which influences the area occupied by expanding *Spartina* clumps. For example, growth of *S. anglica* and *S. maritima* rhizomes is limited by the interference of erosive banks that obstruct their horizontal expansion (Van Hulzen et al., 2007; Castillo et al., 2008b). In contrast, the production of aerial and subterranean biomass is stimulated by high accretion rates for *S. maritima* (Castillo et al., 2008a) and for *S. alterniflora* (Ford et al., 1999; Mendelssohn & Kuhn, 2003; Deng et al., 2008) due to an increase in soil fertility and marsh elevation, reducing nutrient deficiency and flooding stress. In this environment, it is usual to find several levels of rhizome corresponding with different sedimentation events. Thus, hydrologic and sedimentary processes are very important for the inter- and intra-marsh variability of *Spartina* marshes (Montalto & Steenhuis, 2004).

These biotic and abiotic processes usually produce an aggregated spatial pattern of *Spartina* biomass distribution, showing high spatial variability in biomass accumulation (Zedler, 1993). In this context, recorded values of biomass depend very much on the sampling method.

### 2.1 Above-ground biomass

The above-ground biomass of cordgrasses is usually sampled using plots that may be distributed randomly or regularly along radial transects from the edge of the *Spartina* clumps to their centres. Randomly distributed plots are used for extensive and mature prairies where no clumps can be distinguished clearly. If this method is used for a *Spartina* population with isolated clumps, plots should be distributed randomly into the clumps, avoiding non-colonized areas, using stratified sampling. In contrast, radial transects are applied when isolated clumps are easy to distinguish (Castellanos et al., 1994; Nieva et al., 2001a).

The most frequent plot sizes are quadrants of 0.01 m2 (10 cm side) (Buchsbaum et al., 2009; Zhou et al., 2009a), 0.04 $m^2$ (20 cm side) (Castillo et al., 2008a; Culbertson et al., 2008; Buchsbaum et al., 2009; Charles & Dukes, 2009), 0.09 $m^2$ (30 cm side) (Holdredge et al., 2010) and 0.25 $m^2$ (50 cm side) (Schmalzer et al., 1991; Darby & Turner, 2008a,b; Krull & Craft, 2009; Wang et al., 2009; Zhou et al., 2009b) (Table 1).

Larger plots normally show lower biomass values than smaller plots since they include both, areas with high and low shoot densities. However, heterogeneity of data series is usually lower with larger plots. In contrast, smaller plots coinciding with high shoot density areas record higher biomass values, however they may offer very heterogeneous data series when sampling high and low shoot densities. Castillo et al. (2008a) recorded above-ground biomass using quadrants of 0.04 $m^2$ and 1.00 $m^2$ in an expanding population of *S. maritima* and they obtained from two to five times less biomass using the larger plot size. Thus, the spatial scale to analyze cordgrass biomass has to be properly chosen since results are largely determined by it.

The above-ground biomass of cordgrasses also changes temporarily. Thus, higher biomass accumulation values normally coincide with the end of the growing season, which usually matches up with warmer months. For example, higher biomass accumulation in the Southwest Iberian Peninsula has been recorded for *S. maritima* and *S. densiflora* during autumn and the beginning of winter (Castellanos et al., 1994; Nieva et al., 2001). As well, *S. alterniflora* shows higher biomass accumulation at the end of the summer time in invaded Chinese marshes (Zhou et al., 2009a) and in Louisiana estuaries (Darby & Turner, 2008b). Aboveground necromass remains until it is removed by tides, currents or it is decomposed in situ (Schubauer & Hopkinson, 1984; Nieva et al., 2001a). Therefore, biomass records will change depending on the season of the year. More complete sampling would result from recording biomass variations throughout the year. However, recording maximum biomass accumulation at the end of the growing season seems to be a good method for comparing studies (Kirwan et al., 2009).

The above-ground biomass of cordgrasses may also be estimated by allometric relationships relating biomass with shoot density and shoot height (Castillo et al., 2008a; Tyrrell et al., 2008; Gonzalez Trilla et al., 2009). Anyway, when recording *Spartina* above-ground biomass it is important also to document processes that could affect biomass accumulation such as herbivory by vertebrates (i.e. cattle, deers or goose) or invertebrates (i.e. grabs and insects) or mechanical impacts such as tides and currents.

### 2.2 Below-ground biomass

Below-ground biomass studies are much less abundant than those reporting aerial biomass, however subterranean *Spartina* biomass plays very important functions in coastal marshes (Darby & Turner, 2008a,b; Turner et al., 2009) (Table 1).

Most of the roots and rhizomes of cordgrasses are accumulated close to the sediment surface due to their growth-form with superficial rhizomes parallel to the sediment surface and thin

roots, and to anoxic conditions in the sediments, especially at lower elevations, that limit root elongation (Padgett et al., 1998).
The below-ground biomass of *Spartina* species is usually sampled using cores of 20-30 cm long and 10-20 cm diameter that are driven into the sediments and removed containing the soil and the subterranean biomass (roots and rhizomes); 20–50% underestimates can be generated using this method (Johnen & Sauerbeck 1977) (Table 1). Then, cores are usually divided into sections to analyze below-ground biomass distribution with depth (Nieva et al., 2001a; Darby & Turner, 2008a,b; Michel et al., 2009; Zhou et al., 2009a,b). Normally, below-ground biomass is sampled at the same points where the above-ground biomass has been removed previously.
In-growth cores may be used to record below-ground biomass production. In-growth cores are created by removing a determined soil volume, which is replaced with root and rhizome-free sediment collected from an adjacent area. Then, ingrowth cores are removed seasonally, and subterranean biomass is sorted and dried to a constant weight. Total belowground production (g $m^{-2}$ $yr^{-1}$) is calculated by adding together the total amount of live and dead biomass produced at the end of the one year study period (Gallagher et al., 1984; Perry & Mendelssohn, 2009).
Temporal changes in subterranean biomass of cordgrasses are not well established. However, *Spartina* below-ground biomass does not seem to show as clear a temporal pattern as the aerial biomass does, with higher accumulations at the end of the warmer season (Darby & Turner, 2008b).

Fig. 2. Sampling *Spartina versicolor* above-ground biomass along a transect in a representative tussock. Each transect was a belt of contiguous quadrants (10 cm radially × 15 cm wide) across the radius of a tussock; every quadrant sampled a concentric ring around the tussock centre so that the results integrated the zones of different density within the clone.

## 3. Aerial biomass of cordgrasses

*Spartina* species are clonal plants and their ramet distribution and demography is a key factor determining the functioning of these species and their roles in plant communities (Suzuki & Hutchings, 1997). Some cordgrasses grow in dense tussocks (van Groenendael et al., 1996) -"phalanx" growth species after Lovett Doust & Lovett Doust (1982)- and others in sparse clumps - "guerrilla" growth.

Despite the important presence in numerous ecosystems of clonal plants with "phalanx" growth, there have been few detailed studies of their growth strategy (Gatsuk et al., 1980; Bullock et al., 1996; Guardia et al., 2000). *S. argentinensis, S. densiflora, S. bakeri, S. patens* and *S. versicolor* forms dense tussocks that show a high occupation of available space inside tussocks by live shoots in expanding populations and by live and dead shoots in mature populations (Figueroa & Castellanos, 1988; Nieva et al., 2001b). These dense tussocks seem to show a high degree of physiological integration between ramets (Hester et al., 1994 for *S. patens*) reflected in very low mortality rates of young shoots (Pitelka & Ashmun, 1985; Maillette, 1992). Thus, interior areas of "phalanx" tussocks are not easily colonized by other species unless they develop central die-back areas (known as "monk's tonsure"). On the other hand, high stem densities also favor the deposition of particles suspended in the tidal water, an additional source of nutrients (Adam, 1990).

*S. alterniflora, S. anglica, S. foliosa, S. maritima* and *S. x townsendii* are "guerilla" species. They grow in clumps with low shoots densities in comparison with "phalanx" species. These cordgrasses expand with long rhizomes from which a rapid development of aerial tissues is carried out after colonization of bare sediments mainly during warmer periods. Their growth rate seems to decrease when different clumps are close to each other due to higher levels of intraspecific competition between ramets (Castellanos et al., 1998; Li et al., 2009). Intraspecific competition may also inhibite seedling recruitment as has been described for *S. alterniflora* in Willapa Bay (Washington, USA) (Lambrinos & Bandos, 2008).

"Guerrilla" clumps frequently develop central die-back areas throughout their ontogenic development from expanding seedlings to mature tussocks (Turner et al., 2004). Clumps' central areas are more elevated and show less shoot densities than peripherical zones of clumps. Thus, central areas may be colonized by more competitive species that are less tolerant to abiotic stress and that finally displace *Spartina* out by interspecific competition. For example, *S. maritima* facilitates the development of ecological succession by ameliorating anoxia at the centre of its tussocks, so other species such as *Sarcocornia perennis* subspecies *perennis, Sarcocornia perennis x fruticosa* and *Atriplex portulacoides* are able to colonize them and outcompete the small cordgrass (Castellanos et al., 1994; Figueroa et al., 2003).

Most *Spartina* species with a "phalanx" growth-form accumulate more above-ground biomass than cordgrass species growing in "guerrilla". "Guerrilla" species expand faster by rhizomes (e.g. 0.13 m $yr^{-1}$ for *S. maritima* in European marshes following Castillo & Figueroa, 2008) than "phalanx" species (e.g. ca. 0.06 m yr-1 for *S. densiflora* as reported by Nieva et al., 2005 and Kittelson & Boyd, 1997 in European and North American salt marshes, respectively). This faster colonization of surrounded sediment by "guerrilla" species has as a result a lower density of biomass in the occupied space. However, *S. patens* with high ramet densities shows low biomass values compare with *S. alterniflora* with "guerrilla" growth because its shoots are usually much thinner, having smaller leaves (Table 1).

Cordgrass above-ground biomass varies markedly between and within species depending on many different environmental factors and the growth form of each taxon. Minimum aerial biomass values have been recorded for "guerrilla" species growing in low marshes (ca. 100 g DW $m^{-2}$) and maximum values (ca. 15000-30000 g DW $m^{-2}$) have been recorded for "phalanx" species in brackish wetlands (Table 1).

Between "phalanx" growth species, *S. bakeri* accumulates ca. 400-700 g DW $m^{-2}$ in the western Atlantic coast of North America (Chynoweth, 1975; Schmalzer et al., 1991). *S.*

*densiflora* above-ground biomass accumulation changes markedly between different invaded marsh habitats in SW Iberian Peninsula. Thus, dense-flowered cordgrass shows its highest biomass values (higher than 15000 g DW $m^{-2}$) at brackish marshes (Nieva et al., 2001a), with its photosynthetic apparatus being more efficient at lower salinities (Castillo et al., 2005b). In contrast, invading European populations of *S. densiflora* in middle salt marshes show aboveground biomass values ca. 14000 g DW $m^{-2}$ (with ca. 10500 shoot $m^{-2}$). This value falls to ca. 400-600 g DW $m^{-2}$ in low salt marshes (with ca. 6000-7000 shoot $m^{-2}$) (Nieva et al., 2001a; Castillo et al., 2008b) since *S. densiflora* is quite sensitive to long flooding periods (Castillo et al., 2000). In high marshes, *S. densiflora* accumulates intermediated biomass values between low and middle marshes (ca. 6000 g DW $m^{-2}$ with ca. 4000 shoot $m^{-2}$) (Nieva et al., 2001a). Similar above-ground biomass values have been recorded also for South American native populations of *S. densiflora* (Vicari et al., 2002) and invasive populations in California (Moseman-Valtierra et al., 2009) (Table 1). In addition, mature tussocks of *S. densiflora* normally to show elevated wrack accumulation (ca. 800 g DW $m^{-2}$) (Castillo et al., 2008b). Wrack accumulation may interfere with other species colonization of inner areas of *Spartina* clumps.

*S. patens*, another cordgrass with "phalanx" growth, accumulates between 100 and 500 g DW $m^{-2}$ of aerial biomass in North American Atlantic brackish marshes (Silander & Antonovics, 1979; Bertness, 1991; Buchsbaum et al., 2009) and *S. spartinae* accumulates between 200 and 500 g DW $m^{-2}$ in coastal marshes of Texas (MacAtee et al., 1979). *S. versicolor* also forms dense tussocks and colonizes brackish marshes in Europe where it accumulates ca. 3000 g DW $m^{-2}$ (unpublished data) since its shoots are much thicker and taller (ca. 60-95 cm) (Menéndez & Sanmartí, 2007) than those of *S. patens* (ca. 31 cm) (Silander & Antonovics, 1979) (Table 1).

Among cordgrasses with a "guerrilla" growth-form, *S. alterniflora* in western Atlantic low salt marshes accumulates between 100 and 1100 g DW $m^{-2}$. Changes in *S. alterniflora* biomass between populations are related mainly with its short and tall forms, varying its shoot height between 20 and 140 cm, also with a highly variable shoot density that changes markedly between 100 and 4000 shoot $m^{-2}$ (Craft et al., 1999, 2002, 2003; Proffitt et al., 2005; Culbertson et al., 2008; Darby & Turner 2008b; McFarlin et al., 2008; Sala et al., 2008; Tyrrell et al., 2008; Buchsbaum et al., 2009; Gonzalez Trilla et al., 2009; Krull & Craft, 2009; Michel et al., 2009; Holdredge et al., 2010). One year after invading Chinese marshes, *S. alterniflora* accumulated ca. 300-450 g DW $m^{-2}$ with shoot height between ca. 70-240 cm (An et al., 2007), accumulating in mature Chinese populations between 200 and 3700 g DW $m^{-2}$ (Wang et al., 2008; Li & Yang, 2009; Wang et al., 2009; Zhou et al., 2009a,b) (Table 1).

Above-ground biomass of *S. cynosuroides* and *S. maritima* shows similar values than those recorded for *S. alterniflora*, also showing very high inter-population differences. Thus, *S. maritima* accumulates ca. 1000 g DW $m^{-2}$ in low European salt marshes, however it can vary very much depending on the abiotic environment, increasing to ca. 1500 g DW $m^{-2}$ (Benito & Onaindia, 1991; Castellanos et al., 1994; Figueroa et al., 2003; Lillebo et al., 2006; Castillo et al., 2008a,b) (Table 1).

As we have shown above cordgrass above-ground biomass accumulation depends on clone architecture and shoot morphology. These traits show high intraspecific variation as described for *S. alterniflora* (Lessmann et al., 1997; Proffitt et al., 2005), *S. densiflora* (Nieva et al., 2001a; Castillo et al., 2008b), *S. maritima* (Sanchez et al., 1997; Castellanos et al., 1998; Otero et al., 2000; Castillo et al. 2005a; Castillo et al., 2008a,b) and *S. patens* (Silander &

Antonovics, 1979). Intraspecific changes in growth form may be based on phenotypic plasticity or genotypic differences. Thus, several species of *Spartina*, such as *S. alterniflora* or *S. maritima*, show clearly distinguishable tall and short growth forms (Shea et al., 1975; Mendelssohn, 1979; Anderson & Treshow, 1980; Howes et al., 1986; Pezeshki & DeLaune, 1991; Castillo et al., 2005a). Some studies have concluded that the observed variability in growth forms among *Spartina* populations may be the result of genetic differentiation (Gallagher et al., 1988; Sanchez et al., 1997; Proffitt et al., 2003), identifying ecotypes with different canopy heights and biomass accumulation (Lessmann et al., 1997; Daehler, 1999; Otero et al., 2000; Seliskar et al., 2002; Proffitt et al., 2005).

In contrast, other studies have attributed different growth forms to phenotypic plasticity in response to differences in environmental factors (Anderson & Treshow, 1980), such as the availability of nutrients (Dai & Wiegert, 1997; Wigand et al., 2003; Zhao et al., 2010), salinity (Phelger et al., 1971; Trnka & Zedler, 2000) or sediment anoxia (Castillo et al., 2005a). The consequence of this is that the different growth forms are ecophenes. In this context, an increase in shoot height is a commonly reported growth strategy for increased water depth in emergent plants (Grace, 1989; Vandersman et al., 1991; Insausti et al., 2001; Sorrel et al., 2002) including in the genus *Spartina* (Lessmann et al., 1997; Castillo et al., 2005a). Shoots might grow taller in response to sediment anoxia, a response possibly signaled by ethylene (Pezeshki et al., 1993), or in response to increased nutrient mobilization in the sediment under more anoxic conditions (Lenssen et al., 1999). Taller shoots would increase the effective photoperiod (i.e. average leaf emergence from tidal waters during daylight hours), a potent environmental factor in limiting the survival of *Spartina* clumps on low marshes, where a few centimeters of elevation in the tidal gradient may determine the lower distribution limit (Castillo et al., 2000). In addition, taller shoots may play a role in improving the oxygenation of rhizomes and roots via aerenchyma in anoxic environments, as in certain non-tidal wetland species (Sorrel et al., 2002); such internal ventilation might be associated with internal pressurization, as has been described for *S. alterniflora* (Hwang & Morris, 1991). Variations in canopy height of a dominant *Spartina* species can influence ecological functions and the structure of plant and animal communities in a marsh (Seliskar et al. 2002).

As we have reported previously, *Spartina* biomass accumulation depends also on abiotic environmental factors. In warmer locations at lower latitudes, cordgrasses lack a dormant period, sustaining high biomass accumulation rates. For example, created marshes of *S. maritima* in the Odiel Marshes (Southwest Iberian Peninsula) develop faster (within 2-4 years with maximum net aerial primary productivity of ca. 600 g DW $m^{-2}$ $yr^{-1}$ following Castillo et al., 2008a) than North American marshes of *S. alterniflora* (Craft et al., 1999, 2002, 2003; Edwards & Mills, 2005), which seems to be related to warmer winters in Iberian salt marshes. Thus, *S. alterniflora* productivity decreases with latitude and air temperature along the western Atlantic coast of North America (Kirwan et al., 2009). Other climatic factors such as rainfall that determinates erosion, salinity and flooding may also limit cordgrasses biomass accumulation (Gonzalez Trilla et al., 2009). Experimental results in the western Atlantic coast indicated that modest daytime warming increased total above-ground biomass for *S. alterniflora*, but not for *S. patens*. Warming also increased maximum stem heights of *S. alterniflora* and *S. patens* (ca. 8%). In addition, drought markedly increased the total biomass of *S. alterniflora* and the live biomass of *S. patens*, perhaps by alleviating waterlogging of sediments (Charles & Dukes, 2009). On the other hand, *Spartina* biomass also depends on nutrient availability, especially nitrogen, as it has been reported for *S. alterniflora* (Darby & Turner, 2008a; McFarlin et al., 2008).

Biotic direct and indirect interactions also control biomass accumulation of *Spartina* populations. Thus, interspecific competition between two cordgrasses may limit their biomasses. Following the general theory of salt marsh zonation (sensu Pennings & Callaway, 1992 and Pennings et al., 2005): competitive dominants colonize higher elevation in the tidal frame displacing competitive subordinates to more stressful environments with long submergence periods or higher salinities. For example, invasive cordgrass such as *S. alterniflora, S. densiflora* and *S. patens* may displace indigenous cordgrasses (SanLeon et al., 1999; Chen et al., 2004; Castillo et al., 2008b). The outcome of competitive interactions changes depending on the abiotic environment. For example, *S. densiflora* invading European salt marshes displaces the native *S. maritima* at middle and high marshes but it seems to be displaced by small cordgrass at low salt marshes (Castillo et al., 2008b). In this sense, it has been described that the invasion of *S. densiflora* at North American salt marshes is limited by competition with native species (Kittelson & Boyd, 1997) and that *S. patens* competitively excludes *S. alterniflora* and forbs at New England salt marshes (Ewanchuk & Bertness, 2004).

Cordgrass biomass is also affected by competition with other coastal plants as reported along the North-eastern coast of the United States where the reed *Phragmites australis* Cav. is invading high marshes reducing local biodiversity with *S. alterniflora* remaining on the seaward edge of marshes where porewater salinities are highest (Silliman & Bertness, 2004). To the South, in Louisiana, the expansion northward of the tree *Avicennia germinans* (black mangrove) driven by global warming is replacing *S. alterniflora* marshes by mangroves (Perry & Mendelssohn, 2009).

*Spartina* biomass can be also influenced by interactions with marsh fauna. For example, deposit-feeding fiddler crabs (*Uca* sp.) increase *S. alterniflora* biomass accumulation growing on sandy sediment by enhancing nutrient deposition (Holdredge et al., 2010) and grazing by small grazers may carry out a top-down control on *Spartina* biomass dynamic (Sala et al., 2008; Tyrrell et al., 2008).

Above-ground biomass of cordgrasses may collapse very fast as a result of die-back processes related with long flooding periods and sediment anoxia, drought events or nutrient exhaustion (Webb et al., 1995; Castillo et al., 2000; McKee et al., 2004; Ogburn & Alber, 2006; Li et al., 2009). For example, *S. densiflora* invading populations in European salt marshes behave as perennial at middle and high marshes but they are biannual at low marshes. Biannual populations are composed of small tussocks that produce seeds and die, so populations disappear suddenly after two years (Castillo & Figueroa, 2007). *Spartina* shoots are semelparous (they die shortly after their first sexual reproduction event) and their mean shoot life span is about 2 years for species such as *S. densiflora* (Vicari et al., 2002; Nieva et al., 2005) and *S. maritima* (Cooper, 1993; Castellanos et al., 1998). In this sense, some studies predicted that fluctuating environments such as coastal marshes would promote semelparity (Bell, 1980; Goodman, 1984).

On the other hand, cordgrass biomass accumulation is affected negatively, even in the long term, by anthropogenic impacts such as oil spills and erosion (Culbertson et al., 2008), however biomass production may be stimulated by pollutants such as saline oil (Gomes Neto & Costa, 2009).

| *Spartina* Species | Growth form | AGB (g DW m-2) | BGB (g DW m-2) | Location | Sampling method | Source |
|---|---|---|---|---|---|---|
| *S. alterniflora* | *Guerilla* | 469 | | Louisiana, USA | 50 cm quadrants | Hopkinson et al., 1978 |
| | | 137 | - | Oak Island, USA | 24 cm long x 26 cm Ø cores | Ferrell et al., 1984 |
| | | 400-1200 | - | North Carolina coast, USA | 50 cm quadrants | Cornell et al., 2007 |
| | | 100-1100 | - | Great Sippewissett, Massachusetts, USA | 20 cm quadrants | Culbertson et al., 2008 |
| | | - | 150-1200 | Louisiana coast, USA | 50 cm quadrants 30 cm long x 11 cm Ø cores | Darby & Turner 2008a |
| | | 100-900 | 300-2300 | Louisiana coast, USA | 50 cm quadrants 30 cm long x 11 cm Ø cores | Darby & Turner 2008b |
| | | 715-3477 | - | Yangtze River Estuary, China | 25 cm quadrants | Li & Zhang 2008 |
| | | 150 | - | Georgia coast, USA | 50 x 25 cm plots | McFarlin et al., 2008 |
| | | 450-950 | - | Narragansett Bay, USA | 10 cm quadrants | Sala et al., 2008 |
| | | 100-1400 | - | Wells National Estuarine Research Reserve, Maine, USA | Allometric estimation | Tyrrel et al., 2008 |
| | | 1350 | - | Yangtze River estuary, China | 50 cm quadrants | Wang et al., 2008 |
| | | 400 | - | Plum Island Estuary, Massachusetts, USA | 20 cm quadrants | Charles & Dukes, 2009 |
| | | 1400 | - | Altamaha River Mouth, Georgia, USA | 50 cm quadrants | Krull & Craft, 2009 |
| | | - | 6500 | Patuxent River, Maryland, USA | 20 cm long x 16 cm Ø cores | Michel et al., 2009 |
| | | 200 | - | Plum Island Sound, Massachusetts, USA | 10 cm quadrants | Buchsbaum et al., 2009 |

| | | | | | | |
|---|---|---|---|---|---|---|
| | | 200-800 | - | Bahía Blanca Estuary, Argentina | Allometric estimation | Gonzalez Trilla et al., 2009 |
| | | 3700 | - | Yangtze River Delta, China | 40 cm quadrants | Li & Yang, 2009 |
| | | 250-700 | - | Yangtze River Estuary, China | 50 cm quadrants | Wang et al., 2009 |
| | | 700-768 | | Altamaha River, Georgia, USA | 50 cm quadrants | White & Albert, 2009 |
| | | 70-600 | 80-450 | Jiangsu coastland, China | 10 cm quadrants 30 cm deep digging | Zhou et al., 2009a |
| | | 2000 | 4500 | Yancheng Natural Reserve, China | 50 cm quadrants 30 cm deep digging | Zhou et al., 2009b |
| | | 900 | - | Wellfleet, Massachusetts, USA | 30 cm quadrants | Holdredge et al., 2010 |
| ***S. anglica*** | *Guerilla* | 320-1290 | - | Ramalhete marsh, England | 16-19 cm Ø | Neumeier & Amos 2006 |
| ***S. bakeri*** | *Phalanx* | 773 | - | Merritt Island, Florida, USA | 50 cm quadrants | Schmalzer et al., 1991 |
| | | 429 | - | Merritt Island, Florida, USA | 33 cm quadrants | Chynoweth, L.A. 1975 |
| ***S. cynusuroides*** | *Guerilla* | 762-1242 | - | Georgia, USA | | Odum & Fanning, 1973 |
| | | 394 | - | Louisiana, USA | 100 cm quadrants | Hopkinson et al., 1978 |
| | | 840-1080 | | Essex, England | 50 cm quadrants | Potter et al., 1995 |
| | | - | 9400 | Patuxent River, Maryland, USA | 20 cm long x 16 cm Ø cores | Michel et al., 2009 |
| | | 236-832 | - | Altamaha River, Georgia, USA | 50 cm quadrants | White & Albert, 2009 |
| ***S. densiflora*** | *Phalanx* | 400- 15000 | 1000-4500 | Odiel Marshes, SW Iberian Peninsula | 15 x 10 cm plots 20 cm long x 5.5 cm Ø cores | Nieva et al., 2001a |
| | | 475-725 | - | Otamendi Natural Reserve, Argentina | 10 cm quadrants | Vicari et al., 2002 |
| | | 3800-30000 | - | The Tijuana River National Estuarine Research Reserve, California, USA | 50 cm quadrants | Moseman-Valtierra et al., 2009 |

| | | | | | | |
|---|---|---|---|---|---|---|
| ***S. patens*** | *Phalanx* | 900 | - | Louisiana, USA | 56 cm Ø | Hopkinson et al., 1978 |
| | | 400 | - | Plum Island Estuary, Massachussets, USA | 20 cm quadrants | Charles & Dukes, 2009 |
| | | 100-120 | - | Plum Island Sound, Massachussets, USA | 10 cm quadrants | Buchsbaum et al., 2009 |
| ***S. maritima*** | *Guerilla* | 920-930 | - | Ramalhete marsh, England | 16-19 cm Ø | Neumeier & Amos 2006 |
| | | 672-1427 | 1190-8694 | Odiel Marshes, SW Iberian Peninsula | 20 cm quadrants | Castillo et al., 2008a |
| | | 193-486 (T) 1063-4210 (M) | 527-7189 (T) 850-3608 (M) | Tagus (T) and Mondego (M) estuary, Portugal | 30 cm quadrants | Sousa et al., 2008 |
| | | 209-490 | 1510-4268 | Tagus Estuary, Portugal | 30 cm quadrants | Caçador et al., 2009 |
| | | 1085-1313 | - | Mira River, Portugal | 20 cm quadrants | Castro et al., 2009 |
| ***S. spartinae*** | *Phalanx* | 207-513 | - | Texas, USA | 50 cm quadrants | McAtee et al., 1979 |

Table 1. Growth-form ('guerrilla' or 'phalanx' after Lovett Doust & Lovett Doust (1982)) and mean above- and below-ground biomass (AGB and BGB, respectively; in g DW $m^{-2}$) studied location, applied sampling method and source for some cordgrasses species (*Spartina* genus) colonizing coastal marshes.

Fig. 3. Clump of the hybrid *Spartina densiflora x maritima* surrounded by *S. densiflora* and *Sarcocornia fruticosa* in Guadiana Marshes (Southwest Iberian Peninsula).

## 4. Subterranean biomass of cordgrasses

The knowledge of environmental factors determining BGB of cordgrasses is very important for salt marsh conservation and management, as it is a critical factor regulating ecosystem functions. Thus, it seems that it is the plant's belowground accumulation of organic, rather than inorganic, matter that governs the maintenance of mature salt marsh ecosystems in the vertical plane (Turner et al., 2004).

*Spartina* species usually accumulate 2-3 times much more subterranean than aerial biomass. Aerial : the subterranean biomass quotient of cordgrasses is usually lower than 1 (ca. 0.5) (Pont et al., 2002; Windham et al., 2003; Castillo et al., 2008a; Darby & Turner 2008b). Below-ground biomass in cordgrasses carries out very important and diverse functions such as storing of resources in its abundant rhizome system (Suzuki & Stuefer, 1999), fixing the plant to sediments in a very dynamic environment subjected to frequent and intense mechanical impacts (grazing, waves and currents) or exploring the sediments for nutrient uptake. In this sense, competition for nutrients has been identified as a relevant factor organizing salt marsh plant zonation (Brewer, 2003).

As in the case of aerial biomass, the subterranean biomass of cordgrasses varies markedly between and within species. *S. densiflora* accumulates ca. 1000-1600 g DW $m^{-2}$ at low marshes, and ca. 4500-6500 g DW $m^{-2}$ at middle, high and brackish marshes in the SW Iberian Peninsula (Nieva et al., 2001a; Castillo et al., 2008b). Below-ground biomass of *S. versicolor* is ca. 3500 g DW $m^{-2}$ at brackish marshes in the SW Iberian Peninsula (non-published data) (Table 1).

In the Atlantic Coast of North America, *S. alterniflora* growing on sandy sediments accumulates ca. 450 g DW $m^{-2}$ (Holdredge et al., 2010) and ca. 6500 g DW $m^{-2}$ in fine sediments (Michel et al., 2009). In Louisiana salt marshes, Darby & Turner, (2008a,b) reported a below-ground biomass for *S. alterniflora* between 150 and 2300 g DW $m^{-2}$. Subterranean biomass production of *S. alterniflora* in Louisiana salt marshes is about 440 g DW $m^{-2}$ $yr^{-1}$ (Perry & Mendelssohn, 2009) and ca. 4500 g DW $m^{-2}$ in invaded Chinese salt marshes (Zhou et al., 2009b). *S. cynosuroides* accumulates between 760 and 1240 g DW $m^{-2}$ in Georgia and Louisiana marshes (Odum & Fanning, 1973; Hopkinson et al., 1978) and ca. 9400 g DW $m^{-2}$ in high marshes in Maryland, USA (Michel et al., 2009). *S. maritima* accumulates in the sediments between 400 and 8700 g DW $m^{-2}$ at low salt marshes that it usually colonizes (Castellanos et al., 1994; Figueroa et al., 2003; Castillo et al., 2008a; Sousa et al., 2008; Caçador et al., 2009).

*Spartina* below-ground biomass accumulation seemed to be favored by sediment accretion (Castillo et al., 2008a) and cordgrass subterranean biomass influences soil elevation rise by subsurface expansion, organic matter addition and sediment deposit stabilization (Ford et al., 1999; Darby & Turner, 2008a). Sedimentation may also increase the aeration of sediments, favoring root development (Castillo et al., 2008a). Thus, well-drained soils led to more-uniform vertical distribution of BGB for *S. alterniflora* and *S. patens* (Padgett et al., 1998; Saunders et al., 2006).

However, fertilization with nitrogen and phosphorous usually increases *Spartina* above-ground biomass, the addition of these nutrient seems to reduce root and rhizome biomass accumulation (Darby & Turner, 2008a). In view of this result and the importance of subterranean cordgrass biomass for marsh functioning, eutrophication is an important threat to salt marsh conservation.

Fig. 4. *Spartina maritima* prairie, a cordgrass with "guerilla" growth from, starting to be outcompeted by *Sarcocornia perennis* supspecies *perennis* in Odiel Marshes (Southwest Iberian Peninsula).

## 5. Cordgrass biomass and ecosystem functioning

Salt marshes fulfill many functions, such as biodiversity support, water quality improvement, or carbon sequestration and they are floristically simple, often dominated by one or a few herbaceous species (Adam, 1990). In this context, cordgrasses are especially important since they are dominant species in many coastal marshes all around the world.

Cordgrasses are commonly used for salt marsh creation, restoration and protection (Bakker et al., 2002; Fang et al., 2004; Konisky et al., 2006; An et al., 2007; Castillo et al., 2008a; Castillo & Figueroa, 2008). In addition, cordgrasses are also used as biotools for phytoremediation (Czako et al., 2006). Primary productivity and biomass accumulation are important indicators of success for salt marsh creation and restoration projects (Edwards & Mills, 2005). Although plant biomass accumulation is a key factor in the functioning of *Spartina* dominated marshes, other ecological attributes, such as species richness and distribution, benthic infauna density or soil nutrient reservoirs, may develop at different rates than cordgrass biomass in restored wetlands (Craft et al., 1999; Onaindia et al., 2001; Craft et al., 2003; Edwards & Proffitt, 2003).

Below- and above-ground biomasses are key functional traits that play very important roles in the ecological behavior of cordgrasses. Thus, *Spartina* biomass influences on the carbon content of marsh sediments (Tanner et al., 2010), the marsh carbon stock (Wieski et al., 2010), marsh methane emissions (Cheng et al., 2010), salt marsh microbial community (First & Hollibaugh, 2010; Lyons et al., 2010), grazing (Burlakova et al., 2009), sediment dynamic (Neumeier & Ciavola, 2004; Salgueiro & Cacador, 2007; Li & Yang, 2009), etc.

Cordgrass biomass affects the emergent of the habitat structure, facilitating succession development by providing a base for habitat development (Castellanos et al., 1994; Figueroa et al., 2003; Proffitt et al., 2005; Castillo et al., 2008b). For example, *S. maritima* in European low salt marshes, *S. alterniflora* in western Atlantic low salt marshes and *S. foliosa* in Californian low salt marshes are important pioneers and ecosystem autogenic engineers

(Castellanos et al., 1994; Castillo et al., 2000; Proffitt et al., 2005). Thus, sediment deposition develops with the establishment of these foundation cordgrasses at low marshes, which yields abiotic environmental changes such as decreasing anoxia and flooding period (Castellanos et al., 1994; Craft et al., 2003; Bouma et al., 2005; Castillo et al., 2008a; Castillo et al., 2008b).

Fig. 5. Clump of the hybrid of *Spartina foliosa x alterniflora* colonizing a mudflat, where the native *Spartina foliosa* is not able to survive, in San Francisco Bay (California).

On the other hand, biomass production by cordgrasses plays a very important role in the nutrient cycle of coastal marshes. *Spartina* species add organic matter to the sediments that they colonize (Craft et al., 2002; Lillebo et al., 2006) and even to adjacent bare sediments by necromass exportation in the form of dead leaves and shoots (Castillo et al., 2008a).

Although cordgrasses are essential for healthy marsh functioning in their native distribution ranges, some of them are very aggressive when introduce to exotic environments. For example, *S. alterniflora* invades salt marshes in China, Europe and the Pacific coast of North America from the Atlantic coast of America. *S. anglica* is colonizing also Chinese and North American salt marshes coming from European marshes. *S. densiflora* is invading the Pacific coast of Chile and North America, African and European marshes from the Atlantic coast of South America (Bortolus, 2006) where it is a salt-marsh dominant of wide latitudinal range (Isacch et al., 2006). Once introduced by anthropogenic activities, exotic cordgrasses are able to invade contrasted marsh habitats due to their high capacity to colonize as pioneer species new formed environments and disturbed locations, showing a wide tolerance to abiotic stress factors such as salinity, anoxia or long flooding periods (Nieva et al., 1999, 2003; Castillo et al., 2005a). Moreover, *Spartina* species with "phalanx" growth develop very dense tussocks with tall canopy and high above- and bellow-ground biomass, avoiding the colonization of native species, stopping the development of ecological succession during very long periods and representing strong competitors (Figueroa & Castellanos, 1988). In addition, some invasive cordgrasses usually show an abundant seed production and long distance dispersion by tidal water and currents (Kittelson & Boyd, 1997; Nieva et al., 2001a; Castillo et al., 2003; Nieva et al., 2005; for *S. densiflora* in European and North American salt marshes). Alien *Spartina* usually modify the abiotic environment during their invasion faster than native species. For example, the introduced *S. alterniflora* in Chinese salt marshes is significantly more efficient in trapping suspended sediment than the native *Scirpus* and *Phragmites* species (Li & Yang, 2009).

## 6. Conclusions

Cordgrasses usually are dominant species in salt marshes all around the world and they play very important roles in ecosystem functioning. Cordgrass biomass accumulation below and above the sediment surface determines energy and material flows in salt marshes.

Most cordgrasses show markedly spatial variations in their biomass accumulation pattern, depending on biotic and abiotic environmental factors and on their growth form ("guerrilla" versus "phalanx", and "short" versus "tall" form). Thus, specific studies to evaluate the ecological roles of cordgrasses should be carried out for each specific location and for each taxon, analyzing both below- and above-ground biomass production and accumulation. In this context, it is very important to choose an appropriate sampling method adapted to our own goals and that would allow comparisons with previous studies.

Future research is needed specially to improve our knowledge about cordgrass below-ground biomass accumulation, dynamic and functions. The evaluation of the salt marsh ecosystem will be incomplete if based exclusively on what is happening aboveground, or as though what happens aboveground is a satisfactory indicator of what is driving changes belowground. Monitoring programs, for example, could be improved if belowground soil processes were included, rather than excluded, as happens frequently. Furthermore, it may be that because of the dominance of the changes in biomass pools belowground compared to aboveground, what happens belowground may be more influential to the long-term maintenance of the salt marsh than are changes in the aboveground components.

Fig. 6. Salt marsh invaded by the South American neophyte *Spartina densiflora* in Humboldt Bay, California.

Future studies should also analyze specifically the development and functions carried out by recently formed *Spartina* hybrids between native and invasive species invading salt marshes in San Francisco Bay and the South-west Iberian Peninsula. The comparision of the biomass dynamic for these hybrids with their parental species will help us to clarify their ecological roles and to prevent serious environmental impacts.

It is also important to study how invasive cordgrasses respond to intra-specific competition with native species by changing their biomass allocation, accumulation and production. In addition, finding and selecting ecotypes for native cordgrasses with different biomass accumulation patterns would be very usefull to improve our technology for salt marsh restoration projects.

## 7. References

Adam, P. (1990). *Salt marsh Ecology*, Cambridge University Press, 0-521-44823-9, Cambridge

Álvarez, R., Castillo, J.M., Mateos-Naranjo, E., Gandullo, J., Rubio-Casal, A.E., Moreno, F.J. & Figueroa, M.E. (2010). Ecotypic variation in phosphoenolpyruvate carboxilase activity of the cordgrass *Spartina densiflora* throughout its latitudinal distribution range. *Plant Biology*, 12, 1, 154-160, ISSN 1435-8603

An, S.Q., Gu, B.H., Zhou, C.F., Wang, Z.S., Deng, Z.S., Zhi, Y.B., Li, H.L., Chen, L. Yu, D.H. & Liu, Y.H. (2007). *Spartina* invasion in China: implicatiosn for invasive species management and future research. *Weed Research*, 47, 3, 183-191, ISSN 0043-1737

Anderson, C.M. & Treshow, M. (1980). A review of environmental and genetic-factors that affect height in *Spartina alterniflora* Loisel (salt-marsh cordgrass). *Estuaries and Coasts*, 3, 3, 168-176, ISSN 1559-2723

Bakker, J.P., Esselink, P., Dijkema, K.S., van Duin, W.E. & de Jong, D.J. (2002). Restoration of salt marshes in the Netherlands. *Hydrobiologia*, 478, 1, 29-51, ISSN 0018-8158

Bell, G. (1980). The cost of reproduction and their consequences. *American Naturalist*, 116, 1, 45-76, ISSN 0003-0147

Benito, I. & Onaindia, M. (1991). Biomass and aboveground production of 4 angiosperms in Cantabrian (N Spain) salt marshes. *Vegetatio*, 96, 2, 165-175, ISSN 0042-3106

Bertness, M.D. (1991). Zonation of *Spartina patens* and *Spartina alterniflora* in New England salt marsh. *Ecology*, 72, 1, 138-148, ISSN 0012-9658

Bortolus, A. (2006). The austral cordgrass *Spartina densiflora* Brong.: its taxonomy, biogeography and natural history. *Journal of Biogeography*, 33, 1, 158-168, ISSN 0305-0270

Bouma, T.J., De Vries, M.B., Low, E., Peralta, G., Tanczos, C., Van de Koppel, J., Herman & P.M.J. (2005). Trade-offs related to ecosystem engineering: A case study on stiffness of emerging macrophytes. *Ecology*, 86, 8, 2187-2199, ISSN 0012-9658

Brewer, J.S. (2003). Nitrogen addition does not reduce belowground competition in a salt marsh clonal plant community in Mississippi (USA). *Plant Ecology*, 168, 1, 93-106, ISSN 1385-0237

Buchsbaum, R.N., Deegan, L.A., Horowitz, J., Garrita, R.H., Giblin, A.E., Ludlam, J.P. & Shull, D.H. (2009). Effects of regular salt marsh haying on marsh plants, algae, invertebrates and birds at Plum Island Sound, Massachussets. *Wetlands Ecology and Management*, 17, 5, 469–487, ISSN 0923-4861

Bullock, J.M., Silvertown, J. & Clear Hill, B. (1996). Plant demographic reponses to environmental variation: distinguishing between effects on age structure and effects on age-specific vital rates. *Journal of Ecology*, 84, 6, 733-743, ISSN 0022-0477

Burlakova, L.E., Karatayev, A.Y., Padilla, D.K., Cartwright, L.D. & Hollas, D.N. (2009). Wetland restoration and invasive species: apple snail (*Pomacea insularum*) feeding on native and invasive aquatic plants. *Restoration Ecology*, 17, 3, 433-440, ISSN 1061-2971

Caçador, I., Caetano, M., Duarte, B. & Vale, C. (2009). Stock and looses of trace metals froms salt marsh plants. *Marine Environmental Research*, 67, 2, 75-82, ISSN 0141-1136

Caldwell, P.A. (1957). The spatial development of *Spartina* colonies growing without competition. *Annals of Botany*, 21, 2, 203-214, ISSN 0305-7364

Castellanos, E.M., Figueroa, M.E. & Davy, A.J. (1994). Nucleation and facilitation in saltmarsh succession: interactions between *Spartina maritima* and *Arthrocnemum perenne*. *Journal of Ecology*, 82, 2, 239-248, ISSN 0022-0477

Castellanos, E.M., Heredia, C., Figueroa, M.E. & Davy A.J. (1998). Shoot dynamics of *Spartina maritima* in successional and non-successional Mediterranean salt marsh. *Plant Ecology*, 137, 2, 213-225, ISSN 1385-0237

Castillo, J.M., Fernández-Baco, L., Castellanos, E.M., Luque, C.J., Figueroa, M.E. & Davy, A.J. (2000). Lower limits of *Spartina densiflora* and *S. maritima* in a Mediterranean salt marsh determined by different ecophysiological tolerances. *Journal of Ecology*, 88, 5, 801-812, ISSN 0022-0477

Castillo, J.M., Rubio-Casal, A.E., Luque, T., Figueroa, M.E. & Nieva, F.J. (2003). Intratussock shoot distribution and biomass of *Spartina densiflora* Brongn. in an invaded salt marsh. *Lagascalia*, 23, 1, 61-73, ISSN 1047-1991

Castillo, J.M., Redondo, S., Wharmby, C., Figueroa, M.E., Castellanos, E.M., Luque, T. & Davy, A.J. (2005a). Environmental determination of shoot height in populations of the cordgrass *Spartina maritima*. *Estuaries and Coasts*, 28, 5, 761-766, ISSN 1559-2723

Castillo, J.M., Rubio-Casal, A.E., Redondo, S., Álvarez-López, A.A., Luque, T., Luque, C., Nieva, F.J., Castellanos, E.M. & Figueroa, E.M. (2005b). Short-term responses to salinity o fan invasive cordgrass. *Biological Invasions*, 7, 1, 29-35, ISSN 1387-3547

Castillo, J.M. & Figueroa, E. (2007). Effects of abiotic factors on the life span of the invasive cordgrass *Spartina densiflora* and the native *Spartina maritima* at low salt marshes. *Aquatic Ecology*, 43, 1, 51-60, ISSN 1386-2588

Castillo, J.M., Leira-Doce, P., Rubio-Casal, A.E. & Figueroa, E. (2008a). Spatial and temporal variations in aboveground and belowground biomass of *Spartina maritima* (small cordgrass) in created and natural marshes. *Estuarine, Coastal and Shelf Science*, 78, 4, 819-826, ISSN 0272-7714

Castillo, J.M., Mateos-Naranjo, E., Nieva, F.J. & Figueroa, M.E. (2008b). Plant zonation at salt marshes of the endangered cordgrass *Spartina maritima* invaded by *Spartina densiflora*. *Hydrobiologia*, 614, 1, 363-371, ISSN 0018-8158

Castillo, J.M. & Figueroa, E. (2008). Restoring Salt Marshes Using Small Cordgrass, *Spartina maritima*. *Restoration Ecology*, 17, 3, 324-326, ISSN 1061-2971

Castillo, J.M., Ayres, D.R., Leira-Doce, P., Bailey, J., Blum, M., Strong, D.R., Luque, T. & Figueroa, E. (2010). The production of hybrids with high ecological amplitude between exotic *Spartina densiflora* and native *S. maritima* in the Iberian Peninsula. *Diversity and Distributions*, In Press, ISSN 1366-9516.

Charles, H. & Dukes, J.S. (2009). Effects of warming and altered precipitation on plant and nutrient dynamics of a New England salt marsh. *Ecological Applications*, 19, 7, 1758–1773, ISSN 1051-0761

Chen, Z.Y., Li, B., Zhong, Y. & Cheng, J.K. (2004). Local competitive effects of introduced *Spartina alterniflora* on *Scirpus mariqueter* at Dongtan of Chongming Island, the Yangtze River estuary and their potential ecological consequences. *Hydrobiologia*, 528, 1-3, 99-106, ISSN 0018-8158

Cheng, X., Luo, Y., Xu, Q., Lin, G., Zhang, Q., Chen, J. & Li, B. (2010). Seasonal variation in CH4 emission and its 13C-isotopic signature from Spartina alterniflora and Scirpus *mariqueter* soils in an estuarine wetland. *Plant Soil*, 327, 1-2, 85-94, ISSN 0032-079X

Chynoweth, L.A. (1975). Net primary production of *Spartina* and species diversity of associated macroinvertebrates of a semi-impounded salt marsh. *Technical Report* No. 1, NASA, Kennedy Space Center, FL, USA.

Cooper, M.A. (1993). Population biology of *Spartina maritima* and *Spartina anglica* monocultures in estuarine salt marshes. Ph.D. Thesis. University of East Anglia

Cornell, J.A., Craft, C.B. & Megonigal, J.P. (2007). Ecosystem gas exchange across created salt marsh chronosequence. *Wetlands*, 27, 2, 240-250, ISSN 0277-5212

Craft, C., Reader, J., Sacco, J.N. & Broome, S.W. (1999). Twenty-five years of ecosystem development of created *Spartina alterniflora* (Loisel) marshes. *Ecological Applications*, 9, 4, 1405-1419, ISSN 1051-0761

Craft, C., Broome, S. & Campbell, C. (2002). Fifteen years of vegetation and soil development after brackish-water marsh creation. *Restoration Ecology*, 10, 2, 248-258, ISSN 1061-2971

Craft, C., Megonigal, P., Broome, S., Stevenson, J., Freese, R., Cornell, J., Zheng, L. & Sacco, J. (2003). The pace of ecosystem development of created *Spartina alterniflora* marshes. *Ecological Applications*, 13, 5, 1417-1432, ISSN 1051-0761

Culbertson, J.B., Valiela, I., Pickart, M., Peacock, E.E., & Reddy, C.M. (2008). Long-term consequences of residual petroleum on salt marsh grass. *Journal of Applied Ecology*, 45, 4, 1284–1292, ISSN 0021-8901

Czako, M., Feng, X.Z., He, Y.K., Liang, D.L. & Marton, L., (2006). Transgenic *Spartina alterniflora* for phytoremediation. *Environmental Geochemistry and Health*, 28, 1-2, 103-110, ISSN 0269-4042

Daehler, C.C., Anttila, C.K., Ayres, D., Strong, D.R. & Bailey, J.D. (1999). Evolution of a new ecotype of *Spartina alterniflora* (Poaceae) in San Francisco Bay, California, USA. *American Journal of Botany*, 86, 4, 543-546, ISSN 0002-9122

Dai, T. and R. G. Wiegert. 1997. A field study of photosynthetic capacity and its response to nitrogen fertilization in *Spartina alterniflora. Estuarine, Coastal and Shelf Science*, 45, 2, 273-283, ISSN 0272-7714

Darby, F.A. & Turner, R.E. (2008a). Effects of eutrophication on salt marsh root and rhizome biomass accumulation. *Marine Ecology Progress Series*, 363, 63-70, ISSN 0171-8630

Darby, F.A. & Turner, R.E. (2008b). Below- and aboveground *Spartina alterniflora* production in a Louisiana salt marsh. *Estuaries and Coasts*, 31, 1, 223–231, ISSN 1559-2723

Deng, Z., An, S., Zhao, C., Chen, L., Zhou, C., Zhi, Y. & Li, H. (2008). Sediment burial stimulates the growth and propagule production of *Spartina alterniflora* Loisel. *Estuarine, Coastal and Shelf Science*, 76, 4, 818-826, ISSN 0272-7714

Edwards, K.R. & Proffitt, C.E. (2003). Comparison of wetland structural characteristics between created and natural salt marshes in southwest Louisiana, USA. *Wetlands*, 23, 2, 344-356, ISSN 0277-5212

Edwards, K.R. & Mills, K.P. (2005). Aboveground and belowground productivity of *Spartina alterniflora* (smooth cordgrass) in natural and created Louisiana salt marshes. *Estuaries asd Coasts*, 28, 2, 252-265, ISSN 1559-2723

Ewanchuk, P.J. & Bertness, M.D. (2004). Structure and organization of a northern New England salt marsh plant community. *Journal of Ecology*, 92, 1, 72-85, ISSN 0022-0477

Fabre, M.E. (1849).Description d'une nouvelle espèce de *Spartina*, abondante sur une portion du littoral méditerranéen. *Annales des Sciencies Naturelles. Botanique. Paris*, 3, 122-125

Fang, X.B., Subudhi, P.K., Venuto, B.C., Harrison, S.A. & Ryan, A.B. (2004). Influence of flowering phenology on seed production in smooth cordgrass (*Spartina alterniflora* Loisel.). *Aquatic Botany*, 80, 2, 139-151, ISSN 0304-3770

Ferrell, R.E., Seneca, E.D. & Linthurst, R.A. 1984. The effects of crude oil on the growth of Spartina alterniflora Loisel. And Spartina cynosuroides (L.) Roth. *Journal of Experimental Marine Biolgy and Ecology*, 83, 1, 27-39, ISSN 0022-0981

Figueroa, M.E. & Castellanos, M.E. (1988). Vertical structure of *Spartina maritima* and *Spartina densiflora* in Mediterranean Marshes, In *Plant Form and Vegetation Structure*, Werger, M.J.A., van der Aart, P.J.M., During, H.J. & Verboeven, J.T.A. (eds.), 105-108, SPB Academic Publishing,The Hague, The Netherlands.

Figueroa M.E., Castillo J.M., Redondo S., Luque T., Castellanos E.M., Nieva F.J., Luque C.J., Rubio-Casal A.E. & Davy A.J. (2003). Facilitated invasion by hybridization of *Sarcocornia* species in a salt-marsh succesion. *Journal of Ecology*, 91, 4, 616-626, ISSN 0022-0477

First, M.R. & Hollibaugh, J.T. (2010). Environmental factors shaping microbial community structure in salt marsh sediments. *Marine Ecology Progress Series*, 399, 15-26, ISSN 0171-8630

Ford, M. A., Cahoon, D.R. & Lynch, J.C. (1999). Restoring marsh elevation in a rapidly subsiding salt marsh by thin-layer deposition of dredged material. *Ecological Engineering*, 12, 2, 189-205, ISSN 0925-8574

Gallagher, J.L., Wolf, P.L. & Pfeiffer, W.J. (1984). Rhizome and root growth rates and cycles in protein and carbohydrate concentrations in Georgia *Spartina alterniflora* Loisel. plants. *American Journal of Botany*, 71, 2, 165–69, ISSN 0002-9122

Gallagher, J.L., Somers, G.F., Grant, D.M. & Seliskar, D.M. (1988). Persistent differences in 2 forms of *Spartina alterniflora* - a common garden experiment. *Ecology*, 69, 4, 1005-1008, ISSN 0012-9658

Gatsuk, L.E., Smirnova, O.V., Vorontzova, L.I., Zaugolnova, L.B. & Zhukova, L.A. (1980). Age states of plants of various growth forms: a review. *Journal of Ecology*, 68, 2, 675-696, ISSN 0022-0477

Gomes Neto, A. & Costa, C.S.B. (2009). Survival and growth of the dominant marsh grass *Spartina alterniflora* in an oil industry saline waste water. *International Journal of Phytoremediation*, 11, 7, 640–650, ISSN 1522-6514

González Trilla, G., Kandus, P., Negrin, V., Vicari, R. & Marcovecchio, J. (2009). Shoot dynamic and production on a SW Atlantic *Spartina alterniflora* marsh. *Estuarine, Coastal and Shelf Science*, 85, 1, 126–133, ISSN 0272-7714

Goodman, D. (1984). Risk spreading as an adaptive strategy in iteroparous life histories. *Theoretical Population Biology*, 25, 1, 1-20, 0040-5809

Grace, J.B. (1989). Effects of water depth on *Typha latifolia* and *Typha dominguensis*. *American Journal of Botany*, 76, 5, 762-768, ISSN 0002-9122

Guardia, R., Raventos, J. & Caswell, H. (2000). Spatial growth and population dynamics of a perennial tussock grass (*Achnatherum calamagrostis*) in a badland area. *Journal of Ecology*, 88, 6, 950-963, ISSN 0022-0477

Hester, M.W., MCKee, K.L., Burdick, D.M., Koch, M.S., Flynn, K.M., Patterson, S. & Mendelssohn, I.A. (1994). Clonal integration in *Spartina* patens across a nitrogen and salinity gradient. *Canadian Journal of Botany*, 72, 6, 767-770, ISSN 1916-2804

Holdredge, C., Bertness, M.D., Herrmann, N. & Gedan, K. (2010). Fiddler crab control of cordgrass primary production in sandy sediments. *Marine Ecology Progress Series*, 399, 253-259, ISSN 0171-8630

Hopkinson, C.S., Gosselink, J.G. & Parrondo, R.T. (1978). Above ground production of seven marsh plant species in coastal Louisiana. *Ecology*, 59, 4, 760-769, ISSN 0012-9658

Howes, B.L., Dacey J.W.H. & Goehringer, D.D. (1986). Factors controlling the growth form of *Spartina alterniflora* - feedbacks between aboveground production, sediment oxidation, nitrogen and salinity. *Journal of Ecology*, 74, 3, 881-898, ISSN 0022-0477

Hwang, I.H. & Morris, J.T. (1991). Evidence for hygrometric pressurization in the internal gas space of *Spartina alterniflora*. *Plant Physiology*, 96, 1, 166-171, ISSN 0032-0889

Insausti, P., Grimoldi, A.A., Chaneton, E.J. & Vasellati, V. (2001). Flooding induces a suite of adaptive plastic responses in the grass *Paspalum dilatatum*. *New Phytologist*, 152, 2, 291-299, ISSN 0028-646X

Isacch, J. P., Costa, C.S.B., Rodríguez-Gallego, L., Conde, D., Escapa, M., Gagliardini, D.A. & Iribarne, O.O. (2006). Distribution of Salt marsh plant communities associated with environmental factors along a latitudinal gradient on the south-west Atlantic coast. *Journal of Biogeography*, 33, 5, 888–900, ISSN 0305-0270

Johnen, B.G. & Sauerbeck, D.R. (1977). A tracer technique for measuring growth, mass and microbial breakdown of plant roots during vegetation, In: *Soil Organisms as Components of Ecosystems*, Lohm U. & Persson T. (Ed.), 366–373, Ecological Bulletins 25, Stockholm.

Kirwan, M.L., Guntenspergen, G.R. & Morris, J.T. (2009). Latitudinal trends in *Spartina alterniflora* productivity and the response of coastal marshes to global change. *Global Change Biology*, 15, 8, 1982–1989, ISSN 1354-1013

Kittelson, P.M. & Boyd, M.J. (1997). Mechanisms of expansion for an introduced species of cordgrass, *Spartina densiflora*, in Humboldt Bay, California. *Estuaries and Coasts*, 20, 4, 770–778, ISSN 1559-2723

Konisky, R.A., Burdick, D.M., Dionne, M. & Neckles, H.A. (2006). A regional assessment of salt marsh restoration and monitoring in the Gulf of Maine. *Restoration Ecology*, 14, 4, 516-525, ISSN 1061-2971

Krull, K. & Craft, C.B. (2009). Ecosystem development of a sandbar emergent tidal marsh, Altamaha River estuary, Georgia USA. *Wetlands*, 29, 1, 314-322, ISSN 0277-5212

Lambrinos, J.G. & Bandos, K.J. (2008). Habitat modification inhibits conspecific seedling recruitment in populations of an invasive ecosystem engineer. *Biological Invasions*, 10, 5, 729-741, ISSN 1387-3547

Lenssen, J.P.M., Menting, F.B.J., van der Putten, W.H. & Blom, C.W.P.M. (1999). Effects of sediment type and water level on biomass production of wetland plant species. *Aquatic Botany*, 64, 2, 151–165, ISSN 0304-3770

Lessmann, J.M., Mendelssohn, I.A., Hester, M.W. & McKee, K.L. (1997). Population variation in growth response to flooding of three marsh grasses. *Ecological Engineering*, 8, 1, 31-47, ISSN 0925-8574

Li, H. & Zhang, L. (2008). An experimental study on physical controls of an exotic plant *Spartina alterniflora* in Shangai, China. Ecological Engineering, 32, 1, 11-21, ISSN 0925-8574

Li, H. & Yang, S.L. (2009). Trapping effect of tidal marsh vegetation on suspended sediment, Yangtze Delta. *Journal of Coastal Research*, 25, 4, 915-924, ISSN 0749-0208

Li, H., Zhi, Y., An, S., Zhao, L., Zhou, C., Deng, Z. & Gu, S. (2009). Density-dependent effects on the dieback of exotic species *Spartina anglica* in coastal China. *Ecological Engineering*, 35, 4, 544-552, ISSN 0925-8574

Lillebo, A.I., Flindt, M.R., Pardal, M.A. & Marques, J.C. (2006). The effect of *Zostera noltii*, *Spartina maritima* and *Scirpus maritimus* on sediment pore-water profiles in a temperate intertidal estuary. *Hydrobiologia*, 555, 1, 175-183, ISSN 0018-8158

Lovett Doust, L. & Lovett Doust, J. (1982). The battle strategies of plants. *New Scientist*, 95, 81-84, ISSN 0262-4079

Lyons, J.I., Alber, M. & Hollibaugh, J.T. (2010). Ascomycete fungal communities associated with early decaying leaves of *Spartina* spp. from central California estuaries. *Oecologia*, 162, 2, 435-442, ISSN 0029-8549

Maillette, L. (1992). Seasonal model of modular growth in plants. *Journal of Ecology*, 80, 1, 123–130, ISSN 0022-0477

MacAtee, J.W., Scifres, C.J. & Drawe, D.L. (1979). Improvement of Gulf cordgrass range with burning or shredding. *Journal of Range Management*, 32, 5, 372-375, ISSN 0022-409X

McFarlin, C.R., Brewer, J.S., Buck, T.L. & Pennings, S.C. (2008). Impact of Fertilization on a Salt Marsh Food Web in Georgia. *Estuaries and Coasts*, 31, 2, 313–325, ISSN 1559-2723

McKee, K.L., Mendelssohn, I.A. & Materne, M.D. (2004). Acute salt marsh dieback in the Mississippi River deltaic plain: a drought-induced phenomenon? *Global Ecology and Biogeography*, 13, 1, 65-73, ISSN 1466-822X

Mendelssohn, I.A. (1979). Influence of nitrogen level, form, and application method on the growth- response of *Spartina alterniflora* in North Carolina. *Estuaries and Coasts*, 2, 2, 106-112, ISSN 1559-2723

Mendelssohn, I.A., Kuhn, N.L. (2003). Sediment subsidy: effects on soil-plant responses in a rapidly submerging coastal salt marsh. *Ecological Engineering*, 21, 2-3, 115-128, ISSN 0925-8574

Menéndez, M. & Sanmartí, N. (2007). Geratology and decomposition of *Spartina versicolor* in a brackish Mediterrean marsh. *Estuarine, Coastal and Shelf Science*, 74, 1-2, 320-330, ISSN 0272-7714

Michel, J., Nixon, Z., Dahlin, J., Betenbaugh, D., White, M., Burton, D. & Turley, S. (2009). Recovery of interior brackish marshes seven years after the chalk point oil spill. *Marine Pollution Bulletin*, 58, 7, 995–1006, ISSN 0025-326X

Mobberley, D.G. (1956). Taxonomy and distribution of the genus *Spartina*. *Iowa State College Journal of Science*, 30, X, 471-574, ISSN 0038-075X

Montalto, F.A.,& Steenhuis, T.S. (2004). The link between hydrology and restoration of tidal marshes in the New York/New Jersey estuary. *Wetlands*, 24, 2, 414-425, ISSN 0277-5212

Moseman-Valtierra, S.M., Johnson, R., Zhang, R. & Quian, P.Y. (2009). Differences in cordgrass structure between a mature and developing marsh reflect distinct N2-fixing communities. *Wetlands*, 29, 3, 919-930, ISSN 0277-5212

Neumeier, U. & Ciavola, P. (2004). Flow resistance and associated sedimentary processes in a *Spartina maritima* salt-marsh. *Journal of Coastal Research*, 20, 2, 435–447, ISSN 0749-0208

Neumeier, U. & Amos C.L. (2006). The influence of vegetation on turbulence and flow velocities in European salt-marshes. Sedimentology, 53, 2, 259-277, ISSN 0037-0746

Nieva, F.J., Diaz-Espejo, A., Castellanos, E.M. & Figueroa, M.E. (2001a). Field variability of invading populations of *Spartina densiflora* Brong. grown in different habitats of the Odiel marshes (SW Spain). *Estuarine, Coastal and Shelf Science*, 52, 4, 515-527, ISSN 0272-7714

Nieva, F.J.J., Castellanos, E.M. & Figueroa, M.E. (2001b). Effects of light and salinity on seed germination in the marsh invader *Spartina densiflora* Brong., 1829 (Gramineae) from Gulf of Cadiz—Spain. *Boletín de la Real Sociedad Española de Historia Natural*, 96, X, 117–124, ISSN 0366-3272

Nieva, F.J.J., Castellanos, E.M., Castillo, J.M. & Figueroa, M.E. (2005). Clonal growth and shoot demography of the invader cordgrass *Spartina densiflora* Brongn at two contrasting habitats in SW European salt marshes. *Wetlands*, 25, 1, 122-129, ISSN 0277-5212

Odum, E.P. & Fanning, M.E. (1973). Comparision of productivity of *Spartina alterniflora* and *Spartina cynosuroides* in Georgia coastal marshes. *Bulletin of the Georgia Academy of Science*, 31, 1-12

Ogburn, M.B. & Alber, M. (2006). An investigation of salt marsh dieback in Georgia using field transplants. *Estuaries and Coasts*, 29, 1, 54-62, ISSN 1559-2723

Onaindia, M., Albizu, I. & Amezaga, I. (2001). Effect of time on the natural regeneration of salt marsh. *Applied Vegetation Science*, 4, 2, 247-256, ISSN 1402-2001

Otero, X.L., Sánchez, J.M. & Macías, F. (2000). Nutrient status in tall and short forms of *Spartina maritima* in the salt marshes of Ortigueira (NW Iberian Peninsula) as related to physicochemical properties of the soils. *Wetlands*, 20, 2, 461-469, ISSN 0277-5212

Padgett, D.E., Rogerson, C.B. & Hackney, C.T. (1998). Effects of soil drainage on vertical distribution of subsurface tissues in the salt marsh macrophyte *Spartina alterniflora* Lois. *Wetlands*, 18, 1, 35-41, ISSN 0277-5212

Pennings, S.C. & Callaway, R.M. (1992). Salt marsh zonation: the relative importance of competition and physical factors. *Ecology*, 73, 2, 681-690, ISSN 0012-9658

Pennings, S.C., Grant, M.B. & Bertness, M.D. (2005). Plant zonation in low-latitude salt marshes: disentangling the roles of flooding, salinity and competition. *Journal of Ecology*, 93, 1, 159-167, ISSN 0022-0477

Perry, C.L. & Mendelssohn, I.A. (2009). Ecosystem effects of expanding populations of *Avecenia germinans* in a Louisiana salt marsh. *Wetlands*, 29, 1, 396-406, ISSN 0277-5212

Pezeshki, S.R. & DeLaune, R.D. (1991). A comparative-study of aboveground productivity of dominant united-states gulf-coast marsh species. *Journal of Vegetation Science*, 2, 3, 331-338, ISSN 1100-9233

Pezeshki, S.R., Pardue J.H. & DeLaune, R.D. (1993). The influence of soil oxygen deficiency on alcohol-dehydrogenase activity, root porosity, ethylene production and photosynthesis in *Spartina patens*. *Environmental and Experimental Botany*, 33, 4, 565-573, ISSN 0095-8472

Phleger, C.F. (1971). Effect of salinity on growth of a salt marsh grass. *Ecology*, 52, 5, 908-&, ISSN 0012-9658

Pitelka, L.F. & Ashmun, J.W. (1985). Physiology and integration of ramets in clonal plants. In *Population Biology and Evolution of Clonal Organisms*, Jackson, J.B.C., Buss, L.W. & Cook, R.E. (eds.), 399–435, Yale University Press, New Haven, CT, USA.

Pont, D., Day, J.W., Hensel, P., Franquet, E., Torre, F., Rioual, P., Ibanez, C. & Coulet, E. (2002). Response scenarios for the deltaic plain of the Rhone in the face of an acceleration in the rate of sea-level rise with special attention to *Salicornia*-type environments. *Estuaries and Coasts*, 25, 3, 337-358, ISSN 1559-2723

Potter, L., Bingham, M.J., Baker, M.G. & Long, S.P. (1995). The potential of two perennial $C_4$ grasses and perennial $C_4$ sedge as lingo -cellulosic fuel crop in N.W. Europe. Crop establishment and yields in E. Englend. *Annals of Botany*, 76, 5, 513-520, 0305-7364

Proffitt, C.E., Travis, S.E. & Edwards, K.R. (2003). Genotype and elevation influence *Spartina alterniflora* colonization and growth in a created salt marsh. *Ecological Applications*, 13, 1, 180-192, ISSN 1051-0761

Proffitt, C.E., Chiasson, R.L., Owens, A.B., Edwards, K.R. & Travis, S.E. (2005). *Spartina alterniflora* genotype influences facilitation and suppression of high marsh species colonizing an early successional salt marsh. *Journal of Ecology*, 93, 2, 404-416, ISSN 0022-0477

Sala, N.M., Bertness, M.D. & Silliman, B.R. (2008). The dynamics of bottom up and top down control in a New England salt marsh. *Oikos*, 117, 7, 1050-1056, ISSN 0030-1299

Salgueiro, N. & Cacador, I. (2007). Short-term sedimentation in Tagus estuary, Portugal: the influence of salt marsh plants. *Hydrobiologia*, 587, 1, 185-193, ISSN 0018-8158

Sánchez, J.M., Otero X.L., Izco, J. & Macias, F. (1997). Growth form and population density of *Spartina maritima* (Curtis) Fernald in northwest Spain. *Wetlands*, 17, 3, 368-374, ISSN 0277-5212

SanLeon, D.G., Izco, J. & Sanchez, J.M. (1999). *Spartina patens* as a weed in Galician saltmarshes (NW Iberian Peninsula). *Hydrobiologia*, 415, 0, 213-222, ISSN 0018-8158

Saunders, C.J., Megonigal, J.P. & Reynolds, J.F. (2006). Comparison of belowground biomass in C-3- and C-4-dominated mixed communities in a Chesapeake Bay brackish marsh. *Plant and Soil*, 280, 1-2, 305-322, ISSN 0032-079X

Seliskar, D.M., Gallagher, J.L., Burdick, D.M. & Mutz, L.A. (2002). The regulation of ecosystem functions by ecotypic variation in the dominant plant: a *Spartina alterniflora* salt-marsh case study. *Journal of Ecology*, 90, 1, 1-11, ISSN 0022-0477

Schmalzer, P.A., Ross Hinkle, C., & Mailander, J.L. (1991). Changes in community composition and biomassin *Juncus roemerianus* Scheele and *Spartina bakeri* Merr. Marshes on year after a fire. *Wetlands*, 11, 1, 67-86, ISSN 0277-5212

Schubauer, J.P. & Hopkinson, C.S. (1984). Above and belowground emergent macrophyte production and turnover in a coastal marsh ecosystem, Georgia. *Limnology & Oceanography*, 29, 5, 1052–1065, ISSN 0024-3590

Shea, M.L., Warren, R.S. & Niering, W.A. (1975). Biochemical and transplantation studies of growth form of *Spartina alterniflora* on Connecticut salt marshes. *Ecology*, 56, 2, 461-466, ISSN 0012-9658

Silander, J. A. & Antonovics, J. (1979). The genetic basis of the ecological amplitude of *Spartina patens*. I. Morphometric and physiological traits. *Evolution*, 33, 4, 1114–1127, ISSN 0014-3820

Silliman, B.R. & Bertness, M.D. (2004). Shoreline development drives invasion of *Phragmites australis* and the loss of plant diversity on New England salt marshes. *Conservation Biology*, 18, 5, 1424-1434, ISSN 0888-8892

Sorrel, B.K., Tanner, C.C. & Sukias, J.P.S. (2002). Effects of water depth and substrate on growth and morphology of *Eleocharis sphaceolata*: implications for culm support and internal gas transport. *Aquatic Botany*, 73, 2, 93-106, ISSN 0304-3770

Sousa, A. I., Lillelø, A. I., Caçador, I & Pardal, M.A. (2008). Contribution of *Spartina maritime* to the reduction of eutrophication in estuarine systems. *Environmental Pollution*, 156, 3, 628-635, ISSN 0269-7491

Suzuki, J. & Hutchings, M.J. (1997). Interactions between shoots in clonal plants and the effects of stored resources on the structure of shoot populations. In *The ecology and evolution of clonal plants*, de Kroon, H. & van Groenendael, J. (eds.), 311-329, Backhuys Publishers, Leiden, The Netherlands.

Suzuki, J. & Stuefer, J.F. (1999). On the ecological and evolutionary significance of storage in clonal plants. *Plant Species Biology*, 14, 1, 11-17, ISSN 0913-557X

Tanner, B.R., Uhle, M.E., Mora, C.I., Kelley, J.T., Schuneman, P.J., Lane, C.S. & Allen, E.S. (2010). Comparison of bulk and compound-specific d13C analyses and determination of carbon sources to salt marsh sediments using n-alkane distributions (Maine, USA). *Estuarine, Coastal and Shelf Science*, 86, 2, 283-291, ISSN 0272-7714

Thompson, J.D. (1991). Phenotypic plasticity as a component of evolutionary change. *Trends in Ecology and Evolution*, 6, 8, 246-249, ISSN 0169-5347

Thompson, J.D., McNeilly, T. & Gray, A.J. (1991). Population variation in *Spartina anglica* Hubbard,C.E.1. Evidence from a common garden experiment. New Phytologist, 117, 1, 115-128, ISSN 0028-646X

Trnka, S. & Zedler, J.B. (2000). Site conditions, not parental phenotype, determine the height of *Spartina foliosa*. *Estuaries and Coasts*, 23, 4, 572-582, ISSN 1559-2723

Turner, R.E., Swenson, E.M., Milan, C.S., Lee, J.M. & Oswald, T.A. (2004). Below-ground biomass in healthy and impaired salt marshes. *Ecological Research*, 19, 1, 29-35, ISSN 0912-3814

Turner, R.E., Howes, B.L.,Teal, J.M., Milan, C.S., Swenson, E.M. & Goehringer-Toner, D.D. (2009). Salt marshes and eutrophication: An unsustainable outcome. *Limnology and Oceanography*, 54, 5, 1634–1642, ISSN 0024-3590

Tyrrell, M.C., Dionne, M. & Edgerly, J.A. (2008). Physical factors mediate effects of grazing by a non-indigenous snail species on saltmarsh cordgrass (*Spartina alterniflora*) in New England marshes. *ICES Journal of Marine Science*, 65, 5, 746–752, 1054-3139

van Groenendael, J.M., Klimes, L., Klimesová, J. and Hendriks, R.J.J. 1996. Comparative ecology of clonal plants. *Philosophical Transactions of the Royal Society of London*, 351, 1345, 1331-1339, ISSN 0962-8436

Vandersman, A.J.M., Voesenek, L.A.C.J., Blom, C.W.P.M., Harem, F.J.M. & Reuss, J. (1991). The role of ethylene in shoot elongation with respect to survival and seed output of flooded *Rumex maritimus* L plants. *Functional Ecology*, 5, ISSN 304-313.

Van Hulzen, J.B., Van Soelen, J. & Bouma, T.J. (2007). Morphological variation and habitat modification are strongly correlated for the autogenic ecosystem engineer *Spartina anglica* (common cordgrass). *Estuaries and Coasts*, 30, 1, 3-11, ISSN 1559-2723

Vicari, R.L., Fischer, S., Madanes, Bonaventura, N.S. & Pancotto, V. (2002). Shoot population dynamics and production on *Spartina densiflora* (Brong) on the floodplain of the Parana River, Argentina. *Wetlands*, 22, 2, 347-354, ISSN 0277-5212

Wang, J., Zhang, X., Nie, M., Fu, C., Chen, J. & Li, B. (2008). Exotic *Spartina alterniflora* provides compatible habitats for native estuarine crab *Sesarma dehaani* in the Yangtze River estuary. *Ecological Engineering*, 34, 1, 57-64, ISSN 0925-8574

Wang, C.H., Tang, L., Fei, S.F., Wang, J.Q., Gao, Y., Wang, Q., Chen, J.K. & Li, B. (2009). *Ecological engineering*, 3 5, 5, 800–809, ISSN 0925-8574

Webb, E.C., Mendelssohn, I.A. & Wilsey, B.J. (1995). Causes for vegetation dieback in a Louisiana salt marsh: A bioassay approach. *Aquatic Botany*, 51, 3-4, 281-289, ISSN 0304-3770

White, S.N. & Alber, M. 2009. Drought-associated shifts ins *Spartina alterniflora* and *S. cynosuroides* in the Altamaha River estuary. Wetlands, 29, 1, 215-224, ISSN 0277-5212

Więski, K., Guo, H., Craft, C.B. & Pennings, S.C. (2010). Ecosystem functions of tidal fresh, brackish, and salt marshes on the Georgia Coast. *Estuaries and Coasts*, 33, 1,161–169, ISSN 1559-2723

Wigand, C., R. A. McKinney, M. A. Charpentier, M. M. Chintala and G. B. Thursby. 2003. Relationships of nitrogen loadings, residential development, and physical characteristics with plant structure in New England salt marshes. *Estuaries and Coasts*, 26, 6, 1494-1504, ISSN 1559-2723

Windham, L., Weiss, J.S. & Weiss, P. (2003). Uptake and distribution of metals in two dominant salt marsh macrophytes, *Spartina alterniflora* (cordgrass) and *Phragmites australis* (common reed). *Estuarine Coastal and Shelf Science*, 56, 1, 63-72, ISSN 0272-7714

Zhao, Y.J., Qing, H., Zhao, C.J., Zhou, C.F., Zhang, W.G., Xiao, Y. & An, S.Q. (2010). Phenotypic plasticity of *Spartina alterniflora* and *Phragmites australis* in response to nitrogen addition and intraspecific competition. *Hidrobiología*, 637, 1, 143–155, ISSN 0018-5158

Zedler, J.B. (1993). Canopy architecture of natural and planted cordgrass marshes - selecting habitat evaluation criteria. *Ecological Applications*, 3, 1, 123-138, ISSN 1051-0761

Zhou, H.-X., Liu, J. & Qin, P. (2009a). Impacts of an alien species (*Spartina alterniflora*) on the macrobenthos community of Jiangsu coastal inter-tidal ecosystem. *Ecological Engineering*, 35, 4, 521-528, ISSN 0925-8574

Zhou, C., An, S., Deng, Z., Yin, D., Zhi, Y., Sun, Z., Zhao, H., Zhou, L., Fang, C., & Qian C. (2009b). Sulfur storage changed by exotic *Spartina alterniflora* in coastal saltmarshes of China. *Ecological engineering*, 35, 4, 536–543, ISSN 0925-8574

# 2

# Measurement and Assessment Methods of Forest Aboveground Biomass: A Literature Review and the Challenges Ahead

José Návar
*CIIDIR-IPN Unidad Durango*
*Mexico*

## 1. Introduction

The measurement and assessment of aboveground tree biomass (bole, branches, and foliage), or M, plays a key role in the management of forest resources. Estimates are required for evaluating: a) the stocks and fluxes of several biogeochemical elements and b) the amount of primary energy obtainable from forests as an alternative to fossil fuels. Moreover, biomass is a fundamental state variable in several ecological and eco-physiological models (Brown, 1997; Chavé et al., 2005; Návar, 2009a,b; Richardson et al., 2002). The development and use of allometric equations is the standard methodology for the estimation of tree, plot, and regional aboveground biomass (Brown, 1997). Dry weight measurements conducted on harvested trees, fresh and dry weights of biomass components and recording independent tree variables are required to construct allometric equations at the species, stands or tree community levels. Alternate M assessment methods include the multiplication of bole volume by its wood specific gravity; with branch and foliage biomass integrated using other approaches. Standing bole volume, V, can be also multiplied by biomass expansion factors, BEF, at the tree level or stand scale to compute M. Allometric biomass equations can be classified according to the parameter estimation method as empirical, semi-empirical and process, theoretical models. Using three meta-analysis datasets, empirical equations are reported in log-linear (82.6%), non-linear (12.0%), seemingly un-related (3.9%), linear (0.7%), and non-linear seemingly un-related (0.6%) regression. Diameter at breast height, D, and at the bole base, Db, canopy height, H, canopy area, CA, and wood specific gravity, $\rho_w$, are common exogenous variables that individually or in combination explain M with deviations larger than 16% of the mean measured tree aboveground biomass value (Chavé et al., 2005). A fully theoretical, physically parameterized model is available (West et al., 1997), although preliminary evaluations demonstrate that it requires further refinement before can be recommended as a non-destructive M assessment methodology. More flexible, restrictive models that make use of only a small number of harvested trees and fit available allometric equations result in good M approximations (Zianis & Mencuccini, 2004). Semi-empirical non-destructive models based on shape-dimensional analysis and assuming a constant exponent value are being tested for simple and complex forests with compatible preliminary M assessments (Návar, 2010a,b). This wealth of information on biomass allometry necessitates be properly describing, organizing, and classifying in order to better understand weakness and robustness of available methods to compute tree and eventually plot and regional aboveground biomass. For places deprived of tree allometry, a combination of a wide range of allometric equations developed off site appears to improve tree M evaluations according to the Central Limit Theorem.

Biomass stocks and their spatial distribution remain poorly evaluated at the plot scale regardless of the wealth of information on tree biomass allometry (Chavé et al., 2003; Houghton et al., 2001, 20015; Návar et al., 2010). The conventional methodology that expands tree M to sample inventory stands is: a) a grid of sampling plots and b) allometric equations fit tree data recorded in the forest inventory, since there is scarce information on allometric equations that straightforward calculate plot or stand M. New approaches that employ timber volume are named BEF and at the present they require calibration to appraise local plot M (Brown, 2002). Uncertainties of more than two orders of magnitude are identified when calculating plot M by applying different off site allometric models to forest inventory datasets and main sources of variation are: a) the error due to tree measurements, b) ground sampling uncertainty, and above all, c) the error due to the choice of an allometric model relating M to other tree dimensions (Chavé et al., 2003; Návar et al., 2010).

Tree or plot M interpolates at larger spatial scales, AGB, by a variety of field measurements, environmental gradients and remote sensing techniques (Houghton, 2005a,b). A diversity of remote sensing techniques, spatial resolutions, tree and forest attributes, and interpolation methodologies make AGB assessment highly variable, with uncertainties as large as three orders of magnitude. Main sources of variation are attributable to: a) the precision of estimated tree or stand M, b) the interpolation method applied, c) the lack of a good correlation between ground and remote sensing data, d) the correct location of ground data, e) the representativeness of plots across the landscape, f) temporal variations in the satellite image, g) the correct area of each forest class, and h) others. Combining remote field data collection techniques (LIDAR) with locally-derived tree allometry and the semi-empirical shape-dimensional non-destructive model of tree M assessment would eventually improve AGB at the spatial scale of interest.

Given this brief literature review, the reliable M estimation of trees, plots, stands or tree communities remains a key challenge for the successful implementation of sustainable forest management plans. This paper deals with the description of available tree allometry, how they contrast to provide tree, plot and regional M assessments and what are the future challenges ahead. Preliminarily observations point towards the combination of available conventional allometric models with restrictive, semi-empirical and theoretical non-destructive methods of tree or plot M evaluation while universally-applied functions emerge. In addition, the interpolation of improved tree or plot M appraisals to regional scales with a combination of field techniques, environmental gradient approaches and remote sensing methods must eventually improve AGB assessments at regional and national spatial scales.

Key words: Measuring and assessing aboveground biomass, empirical, semi-empirical, theoretic models, tree, stand and regional scales.

## 2. Aboveground tree biomass allometry

### 2.1 Introduction

Aboveground tree and forest biomass is the living and dead matter in standing trees and shrubs and can be classified in foliage, branches, and boles. Bark, hardwood and softwood are timber biomass components. The evaluation of conventional goods and environmental services furnished by forests entails the assessments of tree, stand and regional M. The stocks and fluxes of several biogeochemicals are calculated with the evaluation of M (Brown, 1997; Houghton, 2005). So is the amount of primary energy obtainable from forests as an alternative to fossil fuels (Richardson et al., 2002). In addition, standing aboveground biomass is a fundamental state variable in several ecological and eco-physiological models (Zianis & Mencuccini, 2004).

The development and application of allometric equations is the standard methodology for aboveground tree biomass estimation (Brown et al., 1989; Chavé et al., 2001; 2003; Návar, 2009a). A simple classification of allometric equations based on methods of parameter estimation is: empirical, semi-empirical and theoretical models. Meta-analysis studies report examples of empirical functions (Ter Mikaelian & Korzukhin, 1997; Jenkins et al., 2003; Zianis & Mencuccini, 2004; Zianis et al., 2005; Návar, 2009b). Non-destructive models such as the empirical reductionist (Zianis & Mencuccini, 2004); the semi-empirical shape-dimensional analysis (Návar, 2010a), the constant *B*-slope approach (Návar, 2010b) and process, theoretical methods (West et al., 1999) are also available in the scientific literature.

Empirical allometric equations are statistically parameterized with measured, weighted and recorded field and laboratory tree biomass data. The conventional allometric biomass model ($Ln(M) = Ln(a) + BLn(D) \pm e_i$); where M and D are log transformed and the *a* and *B* the scalar coefficients estimated by least square techniques in linear regression, is the most commonly fitted and reported equation. Other parameter-fitting techniques and mathematical forms of biomass equations are classified as: non-linear, seemingly un-related linear, linear and non-linear seemingly un-related regression, power and exponential functions. Tree diameter recorded at breast height, basal diameter, canopy cover, canopy height and wood specific gravity commonly explain individually or in conjunction tree M with deviations larger than 16% of the mean measured tree M.

Semi-empirical non-destructive methods of tree M computations that focus on independent and easy ways to calculate the conventional allometric scalar coefficients had been recently proposed. They require both physical and statistical parameters. The fractal methodology coupled with shape-dimensional relations was preliminary explored with good degree of precision for temperate trees of northwestern Mexico (Návar, 2010a) and for Mexican tropical forests (Návar et al., 2010). This procedure assumes that bole volume and top height allometric relations suffice to calculate the *a* and *B* scalar coefficients. When contrasted with the conventional allometric model, this method results in compatible tree M assessments. A reduced semi-empirical, non-destructive model that assumes the *B*-scalar exponent is a constant value and the *a*-scalar intercept is a function of the standard wood specific gravity value is also under close mathematical advancement with good preliminary precision for North American temperate trees (Návar, 2010b).

The classic theoretical allometric model, WBE, was developed with the use of fractal techniques (West et al., 1999). Two variables, a C-scalar coefficient and the entire tree specific gravity, $\rho$, suffice to calculate tree M; since it assumes the *B*-scalar exponent is a fixed value of 8/3. The WBE equation is physically parameterized but, at the present, it needs further refinement before can be recommended as non-destructive method of M assessment. Discussions regarding its application are ongoing and they center on the right value of the *B*-slope scalar coefficient that it has been shown to be smaller than 2.67 (Zianis & Mencuccini, 2004, Pilli et al., 2006; Návar, 2009a,b; 2010b).

Other approaches involve the bole volume estimation and then multiplied by the standard wood specific gravity value (Mohren & Klein Goldewijkt, 1990). A dimensionless biomass expansion factor, BEF, escalates bole volume to total tree M (Brown, 1997). Gracia et al. (2004); Lehtonen (2005); Návar-Cháidez (2009); Silva-Arredondo and Návar-Cháidez, (2009) reported independent BEF at the tree level or plot scales, which are developed by employing biomass of the entire tree in conjunction with bole volume allometry.

Tree M assessments are variable regardless of the wealth of information on biomass allometry. For harvested trees, deviations have been reduced to close to 16% with the use of D, H and $\rho_w$, (Chavé et al., 2005). However, the expansion of these equations to trees with other dimensions or outside the forest area where the equation was developed deserves more attention.

### 2.2 The need for tree allometry

A great number of allometric equations have been reported for North American and European tree species and forests (Ter Mikaelian & Korzukhin, 1997; Jenkins et al., 2003; Zianis & Mencuccini, 2004; Zianis et al., 2005; Fehrmann & Klein, 2006; Chojnaky, 2009; Návar, 2009a). Tree allometry for complex tropical (Brown, 1997; Chavé et al., 2001; 2003; 2005) and semi-arid, sub-tropical tree species (Návar et al., 2002a; 2004; Návar, 2009b) and forest plots (Martínes-Yrizar et al., 1992; Návar et al., 2002b) are less represented. At regional scales, current allometric data for complex, diverse tropical forests are almost entirely based on Southeast Asian (Brown, 1997; Ketterings et al., 2001) and South American measurements (Overmann et al., 1994; Araujo et al., 1999; Chavé et al., 2001; 2005; Chambers et al., 2001; Brandeis et al., 2006; Feldpausch et al., 2006). Brown (1997) and Chave et al. (2005) reported a set of allometric equations for tropical world forests; however, several sites were not well typified in this data set. For example, with the exception of the report published by Cairns et al. (2000), most Mexican tropical forests remains with limited information on tree and stand M development, analysis and comparisons.

### 2.3 Development of tree allometry

The development of conventional biomass allometry compels that trees are harvested. Measurements of diameter at breast height and at the bole base are carried out on each standing tree. Top height is better measured once the tree is felled down. Tree dissection into the main biomass components: stem or bole, foliage, and branches are performed on felled trees. Boles are logged into smaller sizes to facilitate weighting. Foliage, branches, and dissected logs are fresh weighted separately per tree. The total fresh weight of each component for each tree is obtained in the field using scales. Samples of each component of each tree are fresh weighted and oven-dried in the laboratory (to constant weight at 70$^0$C). Sample fresh and dry weights must be precisely recorded, since dry to fresh weight ratios for each sample of each component multiplied by the total fresh weight of each biomass component calculate total dry biomass per each tree component. Deviations of this methodology have been proposed where only small portions of each biomass component are weighted and the remaining is calculated by dimensional analysis.

### 2.4 Fitting allometric equations

A data matrix of exogenous, independent variables (D, Db, H, CA, ρw, etc.) and dependent variables (dry foliage, branch, bole, and/or total aboveground biomass) for n, number of harvested trees are available for fitting tree biomass equations. That is, allometry relates one measurement of an organism to another. Easily measured variables such as diameter and top height relates to volume, biomass, etc, which are more difficult to make.

A wide range of empirical allometric models are available in the scientific literature to fit collected biomass data using the independent variables described above. They can be classified as simple log-linear, simple linear, simple non-linear, multiple linear and non-linear, seemingly un-related linear and non-linear regression equations. Power or exponential functions also projects tree M, although they are scarcely reported in the scientific literature. Allometric equations quite often fit each individual biomass component (i.e., see for example the biomass equations compiled by Ter Mikaelian & Korzukhin, 1997; and Návar, 2009b). However, Cunnia and Briggs (1964) showed that when summing the equations for boles, branches, and foliage, results would often deviate from the recorded total aboveground biomass. Therefore, Cunnia & Briggs (1984; 1985) and Parresol (1999; 2001) developed advanced regression techniques and computer programs for estimating coefficient values for endogenous variables that simultaneously calculate individual equation parameters and restrict scalar coefficients to add total tree M. Biomass datasets are also a vital source of information to fit theoretical, semi-empirical non-destructive and restrictive methods of tree M assessment but sometimes other independent variables must be collected.

Biomass datasets should be split into: a) fitting and b) validating models. However, biomass studies are expensive and quite often data is not sufficient to calculate scalar coefficients with small variance that are consistent with population mean parameters. These issues addressed further below must be the center of future allometric studies.

**The Log-linear equation**. The most commonly reported mathematical model for biomass allometry takes the form of the Log linear-transformed function:

$$Ln(M) = Ln(a) + BLn(D) \pm e_i \tag{1}$$

Equation [1] and [2] are similar but not mathematically equivalent:

$$\begin{aligned} Ln(M) &= Ln(a) + BLn(D) \pm e_i \\ M &= \exp(Ln(a) + BLn(D)) \pm e_i \\ M &= aD^B \pm e_i \end{aligned} \tag{2}$$

Where Ln = the logarithmic transformation function; $e_i$ = error.

The scalar coefficients $a$ and $B$ of equations [1] and [2] are calculated by least square techniques in linear regression. Before conducting this statistical test; M and D data is log transformed. The transformation improves parameter estimation by reducing variability and heteroscedasticity. This technique frequently named the intrinsic linear regression entails a weighting parameter to further reduce heterogeneous variance since the logarithmic transformation compresses the data in both axes. When the biomass units are re-transformed back to the original units, the largest data values are often underestimated. Beskersville (1965) recommended to multiply equation [1] by a correction factor, CF, that is calculated as CF = $\exp^{(MSE/2)}$, where: MSE = mean square error of the regression analysis of variance. Equation [1] is the standard, classical allometric biomass model reported in compiled equations by Ter Mikaelian & Korzukhin (1997); Jenkins et al. (2003); Zianis & Mencuccini (2004); and Návar (2009b). The standard error, Sy, of equation [1] is in logarithmic M units and consequently it is not equivalent to: Sy = $\sqrt{MSE}$ where MSE = mean square error. Therefore, equation [1] has to be fitted to the original tree data to evaluate M and with measured and estimated M, Sy can be calculated in conventional M dimensions.

**The Linear equation**. Linear equations frequently reported in allometric studies take the following form:

$$M = a + BX \pm e_i \therefore M = BX \pm e_i \tag{3}$$

Where X = $D^2H$ ($m^3$), BA ($m^2$), Canopy Cover (%).

Least square techniques in linear regression conventionally estimates the scalar coefficients, *a* and *B*. Basal area, the combined variable, $D^2H$, or canopy cover are the explanatory, exogenous variables of equation [3]. The allometric function that entails basal area was originally calculated by measuring M in plots and it has the advantage that can be escalated down to the individual tree level. Examples of this equation are found in Martínez-Yrizar et al. (1992). When using $D^2H$ as independent variable, examples are reported in Padron and Navarro (2004) and in Návar-Cháidez et al. (2004a). Flombaum and Sala (2007) found canopy cover (%) predicted better shrub M for Argentinean semi-arid shrublands. The standard error, Sy, of equation [3] is evaluated in conventional M units.

**The Non Linear equation**. The non-linear equation takes the form of the end portion of model [2], although the error is multiplicative:

$$M = aD^B \cdot e_i \tag{4}$$

Equation [4] is similar but mathematically not equivalent to equation [2], since scalar coefficients are estimated using one of the several non-linear parameter-fitting techniques available such as Newton, Gauss-Newton, Marquardt, etc. That is, scalar coefficient values differ if estimated in linear or non-linear regression techniques. Návar (2009a) reported several examples for temperate tree species of northwestern Mexico. Non-linear models report the analysis of variance in conventional M units and therefore Sy can be straightforwardly computed.

**The multiple linear or non linear equations**. The multiple linear or intrinsically linear equations take the form:

$$M = a + BX_1 + CX_2 + DX_3 + ... + ZX_n \pm e_i \tag{5}$$

Or

$$M = (\exp^{(a+BX_1+CX_2+DX_3+...+ZX_n)}) \pm e_i \tag{6}$$

Or a combination of both. Where: X1, X2, X3,.., Xn = D, $D^2$, $D^2H$, $\rho_w$, $\rho_w D^2H$ ....

Least square techniques in linear or intrinsically linear multiple regression calculates scalar coefficients *a*, *B*, *C*, *D*, *Z*. Brown (1997) and Chavé et al. (2005) reported classical examples for world dry, moist and rain tropical forests that use D, H, and $\rho_w$ as exogenous variables. Multiple linear models supply the standard error of M as the root mean square. Intrinsically linear multiple regression models require a similar procedure to that described in model [1] to calculate Sy in standard M units.

**Seemingly un-related linear regression**. Seemingly un-related regression is the recommended statistical technique to develop tree allometry for endogenous variables, since biomass components are related each other; i.e., leaf biomass relates to branch biomass, these associates to bole biomass, and all these components make total aboveground biomass

(Cunnia & Briggs, 1984; 1985; Parresol, 1999; 2001). Therefore, a simple example of a set of biomass component equations that are linearly related takes the following forms:

$$\begin{aligned}
M &= M_l + M_{br} + M_{bo} \pm e_i \\
M_l &= a_l + B_l(D^2H) \pm e_i \\
M_{br} &= a_{br} + B_{br}(D^2H) \pm e_i \\
M_{bo} &= a_{bo} + B_{bo}(D^2H) \pm e_i \\
M &= a_l + B_l(D^2H) + a_{br} + B_{br}(D^2H) + a_{bo} + B_{bo}(D^2H) \pm e_i = \\
M &= (a_l + a_{br} + a_{bo}) + (B_l + B_{br} + B_{bo})(D^2H) \pm e_i \\
M &= a + B(D^2H) \pm e_i
\end{aligned} \tag{7}$$

Where: l = leaf or foliage, br = branch, bo = bole or stem. All six scalar coefficients, $a_l$, $a_{br}$, $a_{bo}$, $B_l$, $B_{br}$, and $B_{bo}$ are independently and simultaneously estimated by least square techniques in linear regression with the constraint that $M = M_l + M_{br} + M_{bo}$. That is, the sum of each component equals the total tree M.

Parresol (1999) developed and reported this regression technique in computer programs using examples for *P. eliottii* trees. Návar et al. (2004a) fitted this technique for young pine trees of Durango, Mexico and Návar et al. (2004b) did it for semi-arid, sub-tropical shrub species of northeastern Mexico.

**Seemingly un-related non-linear regression**. For un-related non-linear regression, a set of equations written in a simple format are:

$$\begin{aligned}
M &= (M_l + M_{br} + M_{bo}) \cdot e_i \\
M_l &= a_l D^{B_l} \cdot e_i \\
M_{br} &= a_{br} D^{B_{br}} \cdot e_i \\
M_{bo} &= a_{bo} D^{B_{bo}} \cdot e_i \\
M &= (a_l D^{B_l} + a_{br} D^{B_{br}} + a_{bo} D^{B_{bo}}) \cdot e_i
\end{aligned} \tag{8}$$

Where: $M_l$ = foliage biomass, $M_{br}$ = branch biomass; $M_{bo}$ = bole biomass; and *a* and *B* are statistical parameters that are independently and simultaneously estimated by least square techniques in non-linear regression and restricted to provide total aboveground biomass.

Parresol (2001) reported the mathematical development and computer programs for this technique and empirical examples can be found in Návar (2009b).

### 2.5 Examples of empirical equations fitted to an independent dataset

Empirical allometric equations should be cautiously fitted since they may significantly deviate from tree M records. For example, Návar (2009b) reported applications of these empirical equations to a biomass dataset taken for complex semi-arid, sub-tropical shrub species of northeastern Mexico (Figure 1).

All statistical techniques previously described converge into a single equation for more compact biomass datasets as it was shown for young pine trees of northwestern Mexico by Návar (2009b). Note that all equations mimic well the non-linear nature of this M-Db relationship even though multiple linear and seemingly un-related linear equations are fitted to this data.

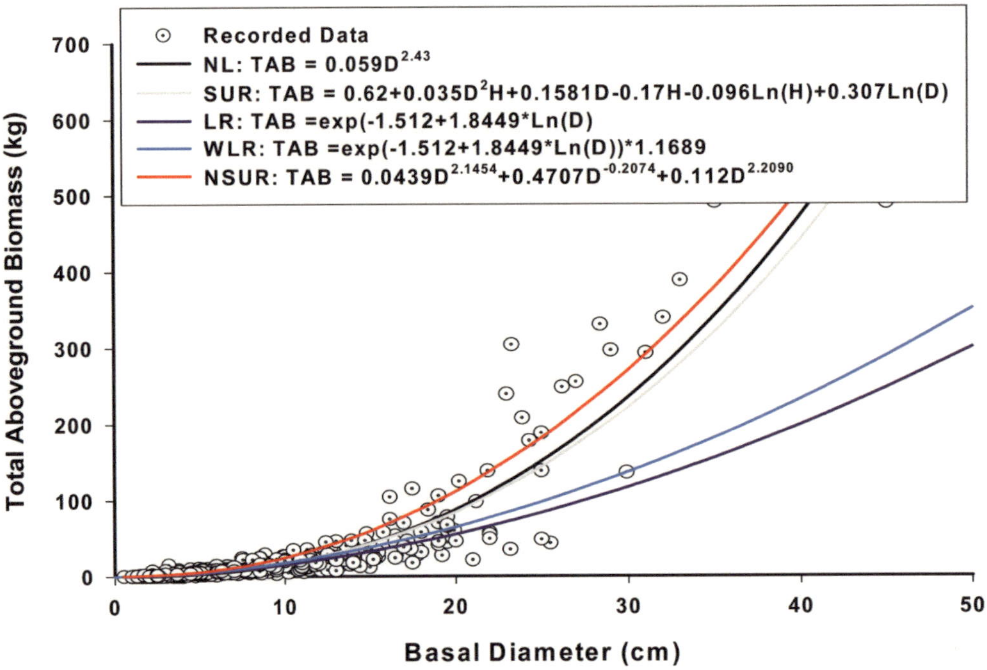


Fig. 1. Five empirical allometric equations fitted to aboveground biomass for 913 shrubs and small trees harvested in northeastern Mexico (NL = Non linear regression, SUR = Seemingly un-related non-linear regression, LR = Linear regression, WLR = weighted linear regression, NSUR = non-linear seemingly un-related regression).

### 2.6 Other allometric models reported in the scientific literature

Other empirical allometric equations reported in the scientific literature were compiled for European tree species by Zianis et al. (2005) but they fall within these major power and exponential classifications: $M=a+bD^c$; $M=a+b[D/(D+cf)]+dX1+\ldots+nXn$; $M=a\cdot(D+1)[b+c\cdot\log(D)]\cdot Hd$; $M=a\cdot(1-\exp(-b\cdot D))^c$.

Where: cf is a standard coefficient; a,b,c are statistical coefficients to be estimated; X1, .. Xn are the independent variables described by D, $D^2$, DH, $DH^2$, etc.

Multicollinearity problems arise when several related exogenous variables explain M making the model unstable in the correct coefficient values.

### 2.7 Examples of the application of empirical tree allometry to biomass data sets

The application of several available allometric equations to independent biomass datasets often results in M assessments with large deviations. Figure 2 shows examples for tropical dry and rain forests as well as for the IAN 710 *Hevea brasiliensis* hybrid trees. Tree allometry is frequently developed with sample data that does not meet the probabilistic sampling requirements. Therefore, local tree allometry improves tree M predictions in contrast to biomass equations developed for tree species with a wide spatial distribution range (Návar-Cháidez, 2010). As a consequence fitting off-site allometric equations often show large tree M uncertainties, which are addressed in the following section of this chapter. Deviations have also been explained by changes in wood specific gravity values and shifts in bole tapering and slenderness. Local, specific tree M allometry has been recommended by Návar-

Cháidez (2010) but further contrasting studies are required in order to understand variations between on-site and off-site equations.

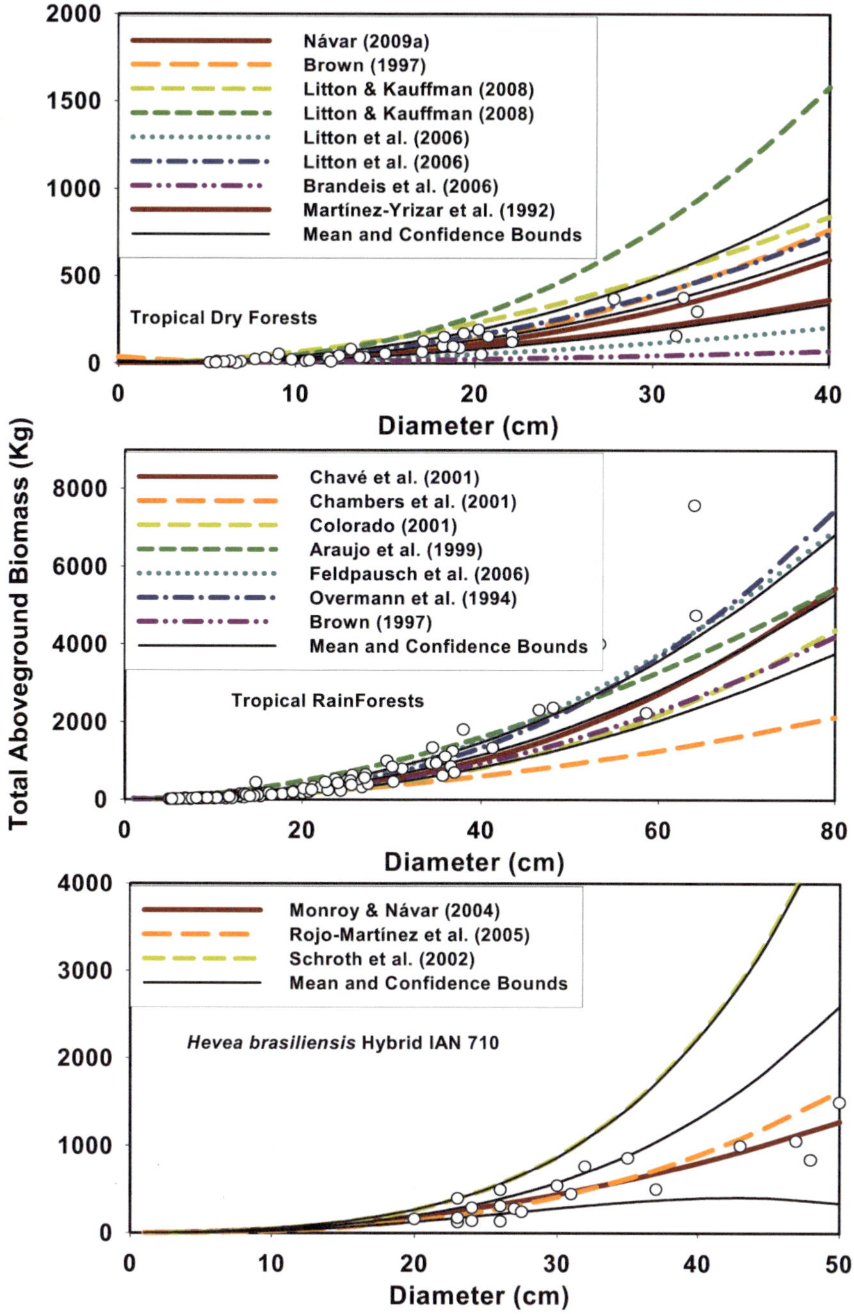


Fig. 2. Aboveground biomass estimates for tropical dry and rain forests as well as for the IAN 710 Hevea brasiliensis Muell. hybrid trees of Mexico (Data Sources: Návar, 2009a; Brown, 1997 and Monroy & Návar, 2005).

### 2.8 Randomness of scalar coefficients with sample size

Empirical allometric equations pose scalar coefficients that are statistically calculated. For example, the conventional model [1] has the *a*–intercept and the *B*-slope scalar values. The scalar coefficients vary with the species, diameter structure, structural tree diversity, the parameter-fitting techniques, sample size, etc. Indeed, scalar parameters show large variations in meta-analysis studies even though they are calculated with the same regression technique. For example, the *B*-scalar exponent has a mean (standard deviation) value of 2.37 (0.27), 2.38 (0.23), and 2.40 (0.22) reported by Zianis & Mencuccini (2004), Návar (2009b), and Fehrmann & Klein (2006), respectively. Návar-Cháidez (2010), in a biomass simulation study with 600 trees, found that the *a* and *B* scalar values randomly oscillate with sample sizes of less than 60 trees pointing out at the need for harvesting sufficient trees to calculate parameter values that are consistent with population means and that have the least variance (Figure 3). Tree M calculations with this set of scalar parameters produce deviations as large as 30%.

### 2.9 Alternate tree allometry models

#### 2.9.1 Reduced number of harvested trees to develop M assessment models

Zianis and Mencuccini (2004) proposed the small tree sampling scheme to simplify allometric analysis irrespective of tree species and forest site. The methodology harvests only two small trees that quite often are D < 25 cm. With recorded D and M, available allometric equations for similar tree species and for similar data ranges found in the forest inventory are fitted. Those equations that predict tree M as close as to the measured values are selected and an average of scalar coefficients *a* and *B* values are estimated to come up with an individual allometric model. This approach was tested with a good balance between acceptable biomass predictions and low data requirements (Návar-Cháidez, 2010). Contrasting results are reported in Figure 6.

#### 2.9.2 The empirical non-destructive model of M assessment

Zianis and Mencuccini (2004) developed de reductionist model [9] by fitting empirical relationships between the *B*-scalar exponent and the slope of the power relationship between H-D, B*. The resulting equation was $B = 1.9262 + 0.6972B^*$; $r^2=0.42$. In a similar fashion, the scalar intercept was projected with the calculated *B*–scalar exponent taking advantage of the good relationship between these coefficients; $a = 7.0281B^{-4.7558}$; $r^2=0.70$. These two equations empirically describe the conventional model [1] scalar coefficients, as follows:

$$M = aD^B$$
$$\left[(7.0281 * B^{-4.7558})D^{(1.9262+0.6972B^*)}\right] \tag{9}$$

Where: B* is the scalar exponent of the H-D power relationship; *B* = is the scalar exponent derived from the empirical equation that relates *B* vs. B*.

#### 2.9.3 The theoretical model of tree M assessment

The West et al. (1999) theoretical model, WBE, was developed using the fractal geometry analysis that applies to natural occurring networks that carry sustaining fluids in organisms, in which each small part of the network is a self-similar replicate of the whole. Hence the fractal model offers much proportionality relating components of structure and function. The WBE framework describes aboveground biomass with the following equation:

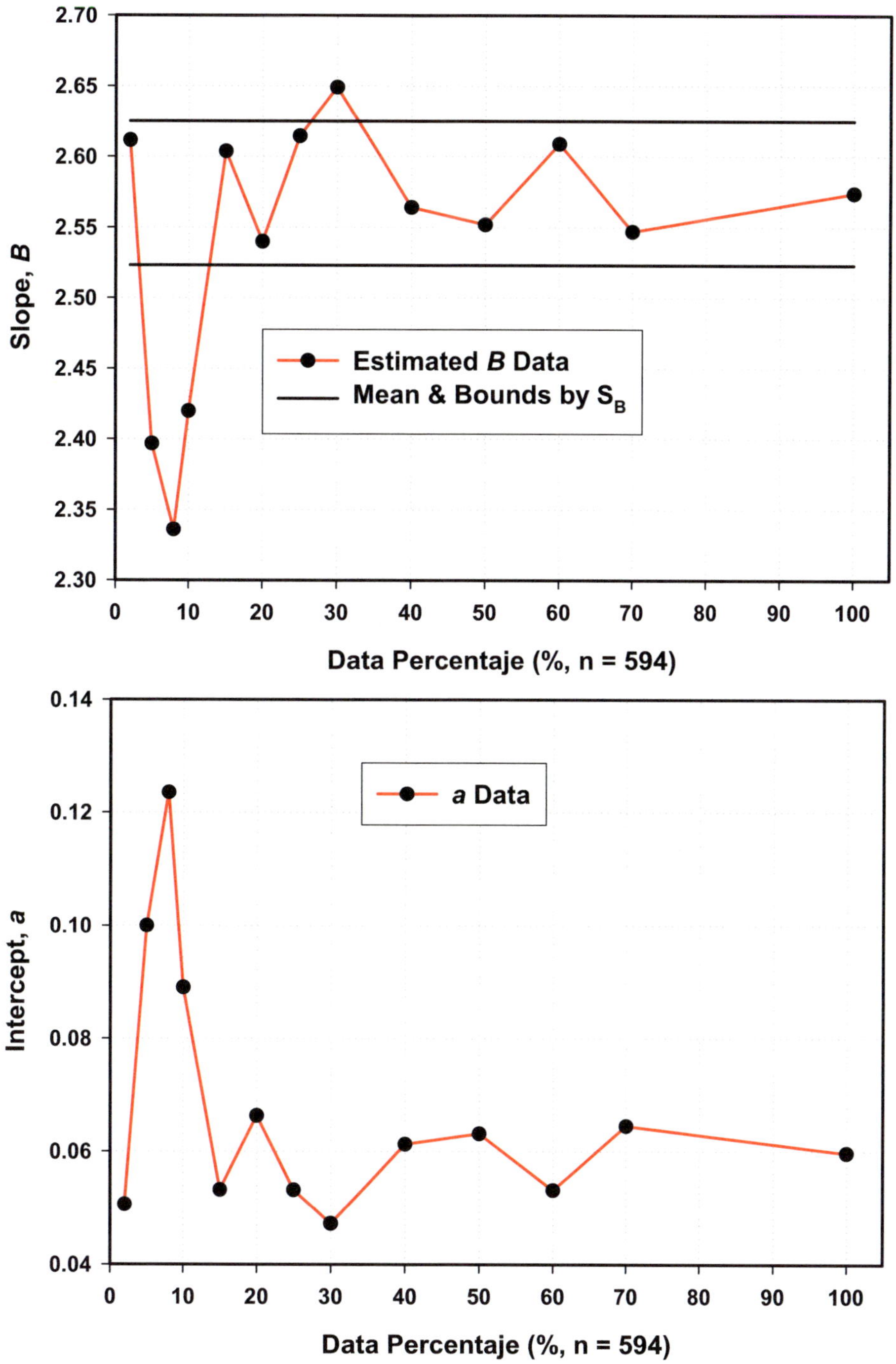


Fig. 3. Randomness of the scalar coefficients *a* and *B* as function of the percentage of data sampled for fitting the conventional allometric model.

$$M = (C\rho)D^{B}$$
$$M = (C\rho)D^{8/3} \quad (10)$$

The scalar exponent, *B*, is fixed to 8/3 (2.67); ρ is the wood specific gravity that is referred as the total tree specific gravity (an average of the specific gravity for wood, bark, branches and leaves); and C = is a proportionality constant. Note that *B* = 2.67.
Comparisons between measurements and predictions by the WBE and other empirical equations were carried out for several biomass data sets. In general, empirical models approximated better recorded tree M values than the WBE one (Zianis and Mencuccini, 2004). Pilli et al. (2006) suggested that M could be estimated by using universal *B* parameters that change with the forest age. Návar (2009b) found evidence that *B* is a function of diameter at breast height and Návar (2010b) successfully tested the hypothesis that *B* is a function of the place where diameter is measured.

### 2.9.4 Semi-empirical non-destructive models of tree M assessment

**a) The shape-dimensional relationships derived from fractal geometry**. Návar (2010a) proposed according to the classical physics equation, that mass is a function of volume x specific gravity. Analogous, the aboveground biomass components are linearly and positively related to stem volume, V, and the entire bole wood specific gravity, ρ; M = (V*ρ). A simple dimensional analysis shows that the volume of a tree bole is $V=(a_vD^2H)$; where $a_v$ = 0.7854 if the bole volume is a perfect cylinder. For temperate tree species of northwestern Mexico, mean $a_v$ values of 0.55 have been calculated demonstrating that tree boles or pieces of stems have a non-standard shape that is only approximated by ideal objects. Therefore, the description of natural items falls beyond the principles provided by Euclidean geometry. Mandelbrot (1983) introduced the neologism of fractal geometry to facilitate the understanding of the form and shape of such objects. A positive number between two and three is a better estimation of the tree's crown dimension, and it is assumed that the overall shape of a tree (stem and crown) may possess a similar fractal dimension. In mathematical terms:

$$V = f(a_v D^d H^h) \quad (11)$$

Where: $a_v$ is a positive number that describes the taper and d and h are positive numbers with $2 \leq d+h \leq 3$.
Since $2 \leq d+h \leq 3$, tree shapes can be described as hybrid objects of surface and volume because they are neither three-dimensional solids, nor two-dimensional photosynthetic surfaces and indentations and gaps are the main characteristics of their structure (Zeide, 1998).
The scaling of H with respect to D has been examined in terms of stress and elastic similarity models following biomechanical principles. When stress-similarity for self-loading dictates the mechanical design of a tree, H is predicted to scale as the ½ power of D (McMahon, 1973) and a final steady state H is attained in old trees that reflects an evolutionary balance between the costs and benefits of stature (King, 1990). Empirical data found that H scales to the 0.535 power of D for a wide range of plant sizes, supporting this hypothesis (Niklas, 1994). However, for local biomass studies, the B* coefficient diverges from the ½ power and it is a function of several variables. Hence, if $H=f(a_hD^{B*})$ with $0<B^*\leq 1 \approx 1/2$, then Eq. (2) becomes

$$V = f(a_v D^d H^h) = (a_v a_h) D^{d+hB^*} \tag{12}$$

Furthermore, if tree biomass is assumed to be proportional to V (with the tree specific gravity as the proportionality constant), then M =f($a_v a_h D^d + H^{B^*} \times \rho$) and in conjunction with Eq. (1), the *B*-scalar exponent, $B_{theoric}$, is:

$$B_{theoric} = d + hB^* \tag{13}$$

And the *a*-scalar intercept, $a_{theoric}$, is:

$$a_{theoric} = (a_v * a_h) \tag{14}$$

Finally, a fully theoretical model that requires the following relationships V = f(D, H) and H = f (D), in addition to the wood specific gravity of the entire aboveground biomass is;

$$M = \rho (a_v a_h) D^{d+hB^*} \tag{15}$$

Model [15] was described as the shape-dimensional analysis approach (Návar, 2010a). In the absence of total aboveground tree $\rho$ and $a_h$ data, the intercept coefficient can be preliminarily derived taking advantage of the good relationship between the scalar coefficients, as follows;

$$a_{theoric} = (a_v * a_h)\rho = f(B_{theoric} = d + hB^*) \tag{16}$$

With this empirical relationship, a final non-destructive semi-empirical model of aboveground biomass assessment is;

$$\begin{aligned} &M = a_{theoric} = (a_v * a_h)\rho = \\ &f(B_{theoric} = d + hB^*) \\ &M = a_{theoric} D^{B_{theoric}} \end{aligned} \tag{17}$$

Meta-analysis studies noted that the scalar coefficients *a* and *B* are negatively related to one another in a power fashion because high values of both *a* and *B* would result in large values of M for large diameters that possibly approach the safety limits imposed by mechanical self loading (Zianis & Mencuccini, 2004; Pilli et al., 2006; and Návar, 2009a; 2009b). This mathematical artifact offers the basic tool for simplifying the allometric analysis of forest biomass in this approach.

In the meantime tree $\rho$ and $a_h$ data is collected, model [17] is a preliminary non-destructive semi-empirical method for assessing M for trees of any size. The procedure can be applied as long as volume allometry is available in addition to the relationship between *a*-*B* that has to be developed preferentially on-site. The methodology is flexible and provides compatible tree M evaluations since large estimated *B* values would have small *a* figures and vice versa. Site-specific allometry can be derived with this model that may improve tree M estimates in contrast to conventional biomass equations developed off-site. Three major disadvantages of this non-destructive approach are: a) the inherent colinearity problems of estimating *a* with *B*, b) the log-relationships between V = f(D, H) and D = f (H) are required in order to estimate *B*, and c) an empirical equation that relates *a* to *B* should be developed on site or alternatively use preliminary reported functions by Zianis & Mencuccini (2004) and by Návar (2009a; 2010a). All these three equations estimate compatible *a*-intercept values with an estimated $B$ slope coefficient. Examples of the application of this semi-empirical model are reported in Figure 6.

**b) Reducing the dimensionality of the conventional allometric equation by assuming a constant $B$ slope coefficient value**. The development of a model that is consistent with the WBE (model [10]) and the conventional log-transformed, most popular equation (model [1]) was proposed by Návar (2010b). Models [1] and [10] have the following common properties: $a = C\rho$; $B_{WBE} \neq B$; $B_{WBE} = 2.67$ and $B$ is a variable that it is a function of several tree and forest attributes, including sample size; they both feed on diameter at breast height as the only independent variable. The main WBE model assumption is that the $B_{WBE}$-scalar slope coefficient is a constant value. This assumption has spurred recent research on semi-empirical allometric models. Hence, Ketterings et al. (2003) and Chavé et al. (2005) reduced the dimensionality of model [1] by proposing a constant $B$-slope coefficient, as well. Tree geometry analysis and assuming that D scales to 2.0H; where H is the slope value of the H = f(D) relationship; i.e., $D^{2.0H}$ are some methods justified for finding this constant. In this report, I hypothesized, according to the Central Limit Theorem, that compilations and Meta analysis studies on biomass equations should shed light onto the population mean $B$-scalar slope coefficient value.

Návar (2010b) summarized several Meta analysis studies on aboveground biomass. Table 1 shows statistical results of these studies compiled from the work conducted by Jenkins et al. (2003); Zianis and Mencuccini (2004); Pilli et al. (2006); Fehrmann and Kleinn (2006); Návar (2009a,b) where there is increasing evidence that the population mean $B$-value is around 2.38. This coefficient differs from the WBE scaling exponent. The Návar (2010b) equation,

| | | Scalar coefficients | | | | | | | | |
|---|---|---|---|---|---|---|---|---|---|---|
| | | $a$ | | | $a$-re-calculated | | | $B$ | | |
| | N | $\bar{x}$ | σ | CI | $\bar{x}$ | σ | CI | $\bar{x}$ | σ | CI |
| Jenkins et al. (2003) | 10(2456) | 0.11 | 0.03 | 0.02 | 0.12 | 0.03 | 0.02 | 2.40 | 0.07 | 0.05 |
| Ter Mikaelian and Korzukhin (1997) | 41 | 0.15 | 0.08 | 0.03 | 0.11 | 0.04 | 0.01 | 2.33 | 0.17 | 0.05 |
| Fehrmann and Klein (2006) | 28 | 0.17 | 0.16 | 0.06 | 0.12 | 0.02 | 0.01 | 2.40 | 0.25 | 0.09 |
| Návar (2009b) | 78 | 0.16 | 0.15 | 0.03 | 0.14 | 0.09 | 0.02 | 2.38 | 0.23 | 0.05 |
| Návar (2010a) | 34 | 0.10 | 0.11 | 0.04 | 0.12 | 0.05 | 0.02 | 2.42 | 0.25 | 0.08 |
| Zianis and Mencuccini (2004) | 277 | 0.15 | 0.13 | 0.01 | 0.12 | 0.04 | 0.01 | 2.37 | 0.28 | 0.03 |
| $\mu$ | | 0.14 | 0.11 | 0.03 | 0.12 | 0.05 | 0.01 | 2.38 | 0.21 | 0.06 |

N = number of biomass equations; $\bar{x}$ = average coefficient value; σ = Standard deviation; CI = confidence interval values (α = 0.05; D.F = n-1); $\mu$ = population mean. Jenkins et al. (2003) compiled 2456 grouped in 10 biomass equations for temperate North American clusters of tree species. Ter Mikaelian and Korzukhin (1997) reported equations for 67 North American tree species but I employed only 41 equations that describe total aboveground biomass. Návar (2009b) reported a Meta-analysis for 229 allometric equations for Latin American tree species but only 78 fitted the conventional model for aboveground biomass. Návar (2010) reported $B$-scalar exponent values for 34 biomass equations calculated from shape-dimensional analysis. Zianis and Mencuccini (2004) reported equations for 279 worldwide species. It is recognized that several studies report the equations that were compiled by Jenkins et al. (2003).

Table 1. Scalar coefficients of the allometric conventional model and re-calculated $a$-scalar intercept values assuming that $B$ = 2.38 for six meta-analysis studies.

consistent with the work conducted by Burrows (2000) and Fehrmann and Kleinn (2006), shows that the scaling exponent of the WBE model is correct as long as $D_{0.10}$ m is reported in the allometric model. Enquist et al. (1998) and West et al. (1999) defined that the WBE approach was derived on the assumption that the relationship between diameter and tree height, H, scales with the assumed exponent value of 2/3. This coefficient has been found to be close to ½ as it was discussed above.

The assumption of a constant *B*-scaling exponent value necessitates the re-calculation of the *a*-scalar intercept value for available allometric equations. A graphical example for this approach is shown in Figure 4 for 41 total aboveground biomass equations reported by Ter Mikaelian & Korzukhin (1997).

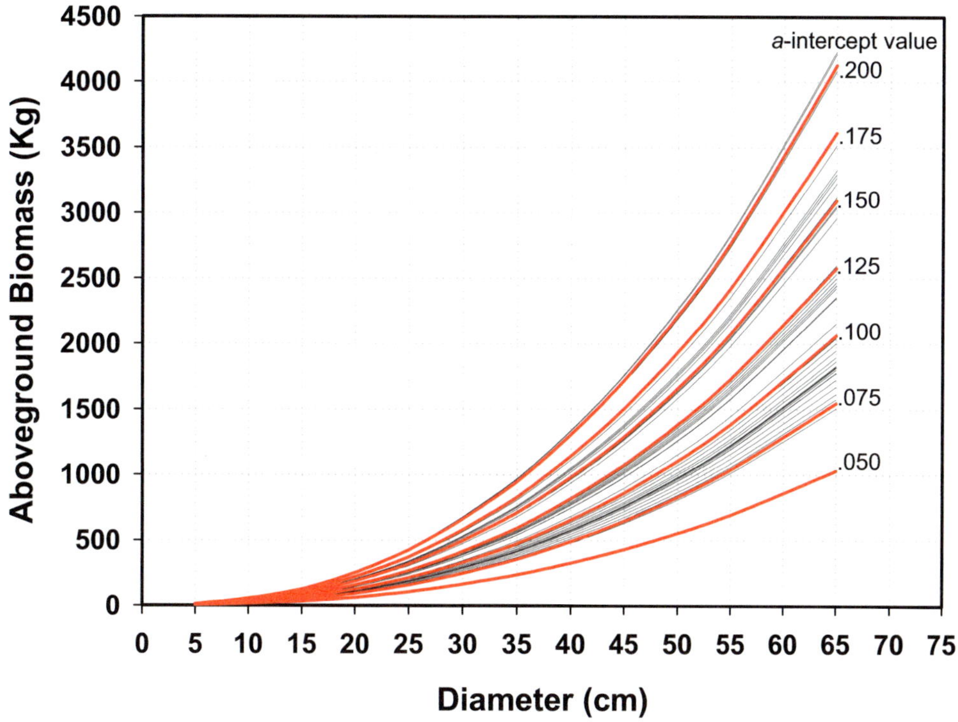


Fig. 4. Total aboveground biomass equations for 41 North American tree species reported by Ter Mikaelian and Korzukhin (1997) overlapped with allometric equations that assume a constant *B*-slope value of 2.38 and re-calculating the *a*-intercept scalar coefficients. Note the suitability of the reduced semi-empirical, non-destructive model of tree M assessment.

The re-calculation of the *a*-intercept is not straightforward. That is, the mathematical solution for the *a*-scaling intercept is not unique. For a reported biomass equation, it is a function of D, as it is described in the following example:

$$M = a_{kn} D^{b_{kn}} \therefore M = a_{ukn} D^{2.38}$$
$$a_{ukn} = \frac{a_{kn} D^{b_{kn}}}{D^{2.38}} = a_{kn} D^{(b_{kn} - 2.38)} \quad (18)$$

Using the example for the *Alnus rugosa*, Ter Mikaelian and Korzukhin (1997) reported the following equation: Ln(M) = 0.2612+2.2087Ln(D). Then, by assuming that the *B*-scalar exponent value is 2.38 instead of 2.2087, the new $a_{unk}$-intercept figure is mathematically solved as follows:

$$\begin{aligned} a_{unk} &= \frac{0.2612D^{2.2087}}{D^{2.38}} = \\ &= 0.2612D^{(2.2087-2.38)} \\ &if\, D = 10\,cm; a_{unk} = 0.1760 \\ &if\, D = 70\,cm; a_{unk} = 0.1261 \end{aligned} \tag{19}$$

Using simulated M-D data, the statistical $a_{unk}$-intercept value would be 0.1229. Therefore, the mathematical method of finding the value of $a_{unk}$ is skewed. In the absence of a statistical program, it is recommended to estimate the *a*-scaling intercept by mathematically solving equation [19] with the largest D value recorded in the biomass study or in the forest inventory. The re-calculation of the *a*-scalar intercept can also be derived with the assumption that $B$ = 2.67 or any other *B*-constant coefficient and produce similar goodness of fit. For 41 allometric aboveground equations reported by Ter Mikaelian and Korzukhin (1994), the mean (confidence interval) *a*-scalar intercept value is 0.1458 (0.026). Re-calculated values with the assumption that $B$ = 2.38 and that $B$ = 2.67 result in mean values of 0.1174 (0.012) and 0.042 (0.0045), respectively. The recalculated *a*-value with the assumption that $B$ = 2.38 outcome consistent and unbiased *a*-intercept figures, statistically similar to those of the original equations (Table 1). The assumption that $B$ = 2.67 deviates notoriously the intercept coefficient values by 3.5 orders of magnitude. That is, the WBE model has to be re-defined in either the *B*-scalar exponent to 2.38 or the C coefficient to a higher value.

A set of biomass equations would have a constant *B*-scalar exponent, a set of re-calculated $a_{unk}$ figures and standard $\rho_w$ values, a data source sufficient to construct the reduced semi-empirical, non-destructive method of M assessment. This methodology assumes: a) that the bole wood specific gravity, $\rho_w$, is similar to the entire tree specific gravity, $\rho$, value; and b) that $a_{unk}$ and $\rho_w$ are linearly related with a 0 intercept, and a slope coefficient that describes the C proportionality constant of the WBE model. Návar (2010b) derived the following relationship: $M = (0.2457(\pm 0.0152))\rho_w{*}D^{2.38}$ for 39 biomass equations for temperate North American tree species. That is, the equation within brackets computes the *a*-scalar intercept with only $\rho_w$ values. This mathematical function is called the Návar (2010b) reduced equation and it is expected to vary between forests and between forest stands. Therefore, this relationship must be locally developed when information is available. Brown (1997) and Chavé et al. (2005) for worldwide tropical species and Miles and Smith (2009) for North American tree species reported comprehensive lists of $\rho_w$ values. If for one moment, it is again assumed that $\rho_w = \rho$, and that $B$ = 2.38, then the C coefficient of the Návar (2010) model would have confidence bounds of 0.2304 and 0.2609 for North American temperate tree species, respectively. The application of this model to 10 clusters of species reported by Jenkins et al., (2003) is reported in Figure 5. The Návar (2010b) reduced model deviates notoriously for the woodland tree species showing that it is specific in nature.

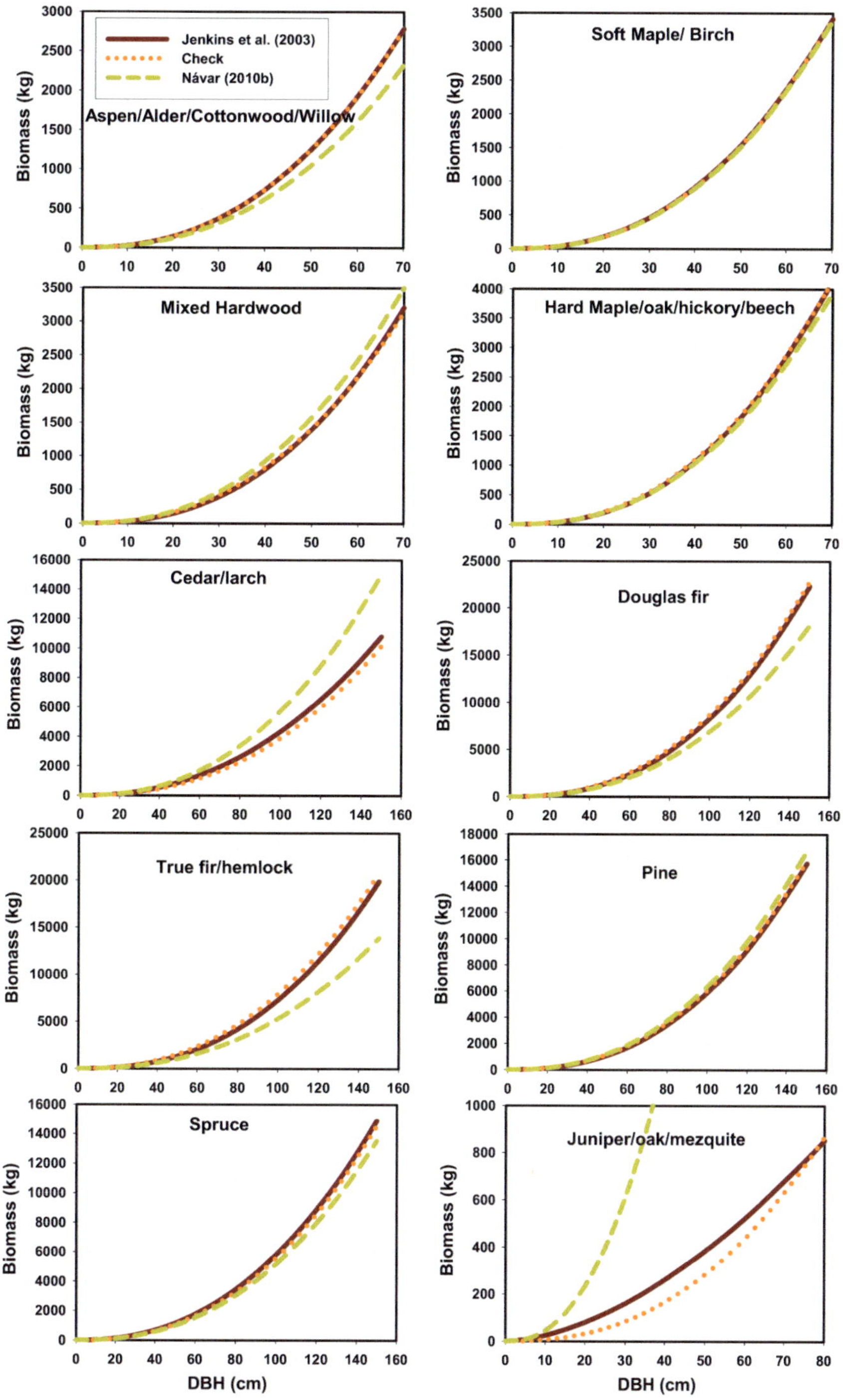


Fig. 5. Contrasts between the reduced semi-empirical, non-destructive model of Návar (2010b) and empirical equations for 10 clusters of tree species reported by Jenkins et al., (2003) for North American tree species.

## 2.10 Examples of semi-empirical methods of tree M assessment

Projected tree M values by the restrictive, the reduced, and the shape-dimensional non-destructive, semi-empirical models reside within the confidence bounds of the conventional model for most allometric equations tested for northern Mexico (Figure 6). The shape-dimensional non-destructive model proposed by Návar (2010a) fits better biomass datasets than the reduced or the restrictive models, since the later model estimates the *a*-intercept coefficient with an $r^2 = 0.65$. Equations reported to estimate *a* with *B*, instead of with $\rho_w$ has an $r^2$ value > than 0.70 (Zianis and Mencuccini, 2004; Fehrmann and Klein, 2006; Pilli et al., 2006; Návar, 2010a). Indeed, Návar (2010a), in a simulation study, observed that $r^2 > 0.90$ for relationships derived for temperate tree species of northwestern Mexico.

The reduced non-destructive, semi-empirical model reported in here can be additionally employed for checking the consistency of available conventional allometric models. That is, equations that trespass *a*-intercept lines would biased M estimates. The limits of most empirical allometric equations can be easily determined using this non-destructive approach. The limits of biomass equations can be found just before they trespass a lines. Hence, this technique is handy for finding the right equations, their limits and as a consequence M estimates would be improved for any forest.

## 2.11 Future directions in the development of semi-empirical methods of M assessment

The tendency of semi-empirical and theoretical process studies to derive constant values that easily describe the mass of trees has become the center of current allometric studies. The methodology proposed by the theoretical and semi-empirical models is the basis for further development and improvement of mixed, process models. Full process models that deterministically assess tree M could never be developed since the variance in aboveground biomass data is hard to be fully explained by conventionally measured tree variables. Therefore, the need for semi-empirical techniques that convey physiological basis such as those proposed by West et al. (1999) and by Návar (2010a,b) derived from fractals, reduced and shape-dimensional analysis.

The empiricism of any non-destructive techniques of tree M assessment would arise early in the bole volume estimation. For example, the Schumacher and Hall (1933) allometric bole volume equation, i.e., $Ln(V) = Ln(a_v) + dLn(D) + hLn(H)$; $a_v D^d H^h$, may have also constant d and h scaling exponents for most trees and the $a_v$ intercept scaling coefficient varies within trees and in trees between forests. If so, the $a_v$ intercept scaling coefficient of the Schumacher and Hall (1933) volume equation would improve the description of the third dimension of timber by incorporating its shape that is intrinsically related to the taper. Just as the *a*-scalar intercept coefficient of the allometric biomass equations describe the fourth dimension of timber, its ρ, the h scaling exponent partially explains the first dimension of timber, its slenderness. These arguments physically suggest that M of a tree with diameter recorded at breast height, D, should be proportional to the product of ρ times volume (V), and that volume is a function of basal area x height; as follows:

$$M = V\rho \tag{20}$$

When model [20] is further developed by coupling the Schumacher and Hall (1933) volume equation and the power function that relates H to D it would result in model [15].

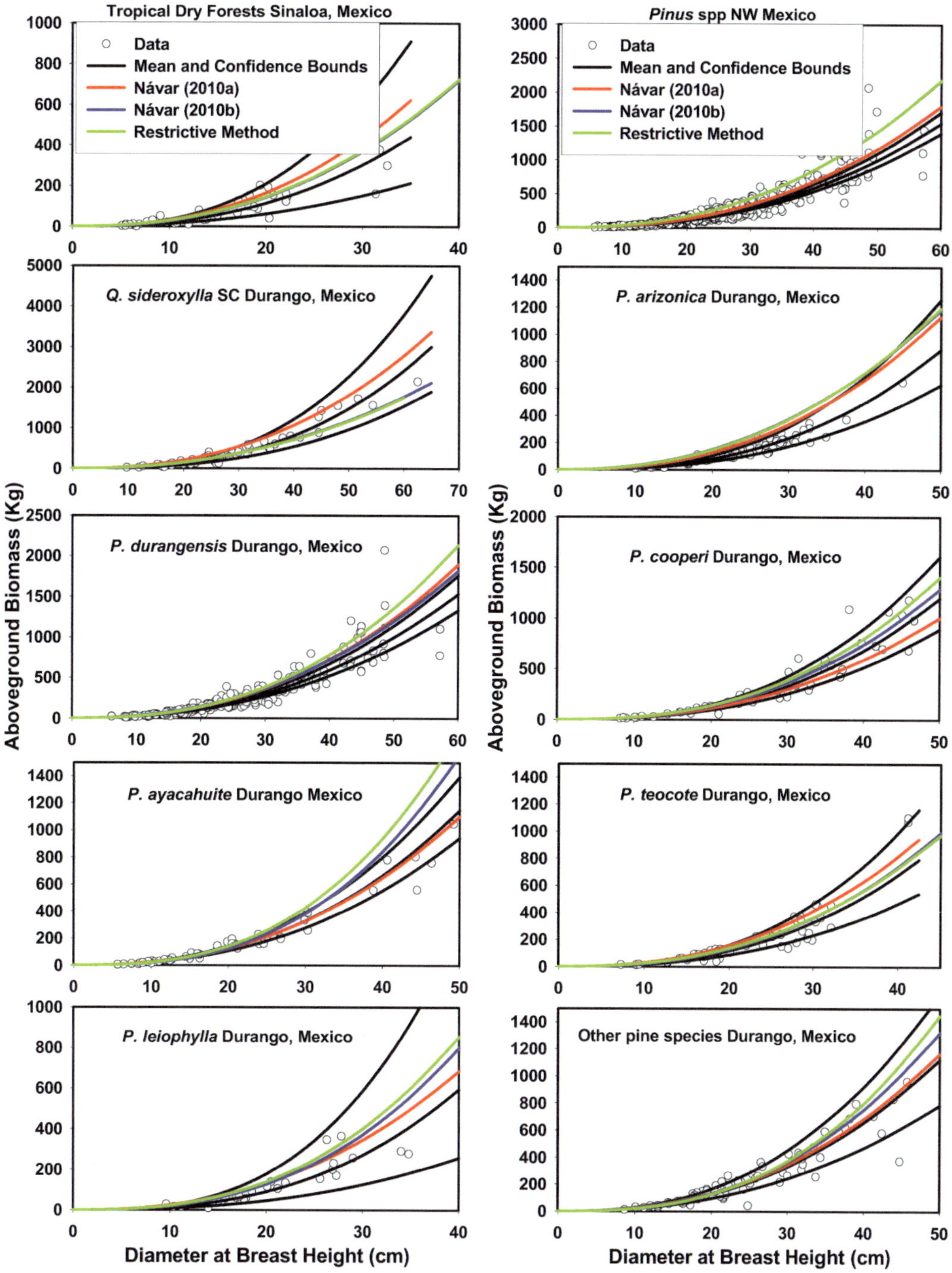


Fig. 6. Testing the restrictive, the reduced and the shape-dimensional, semi-empirical, non-destructive model performance for 10 independent allometric studies collected from northwestern Mexico. The regression lines, raw data and confidence bands on the *B*-value of the conventional allometric model are also depicted.

Equation [15] is similar to the theoretical WBE model by assuming that C = ($a_v$ x $a_h$) and that $B$ = 2.67 = d+hH*. Empirical contrasts of the $B$-scalar exponent values calculated from shape-dimensional analysis and the constant value of the WBE model show that they are statistically different for 34 allometric studies conducted in northern Mexico. The semi-empirical non-destructive model [15] is not different either to those equations proposed by Chave et al. (2005) or by Ketterings et al. (2001), which are reported as models [21] and [22], respectively.

$$M = \rho(C) \cdot D^{2+H^*} \tag{21}$$

$$M = \rho_w(0.11) \cdot D^{2+H^*} \tag{22}$$

Where: H* is the scaling exponent of the power function of the H-D relationship and C is a proportionality constant. Note that Ketterings et al. (2001) proposed that C = 0.11 for tropical trees of south East Asia. The C coefficient values calculated by Návar (2010d) are different than the one proposed by Ketterings et al. (2001), since it had a mean (confidence bound) value of 0.2457 (±0.0152) for North American tree species.

The $B$-scalar exponent 2+H* reported in equations [21] and [22] differs from the empirical value noted in meta-analysis and shape-dimensional studies as 2.38 by Návar (2010b) and the exponent coefficient proposed by West et al. (1999) as 2.67. The H* coefficient has an approximate mean value of 0.53 (McMahon, 1973; Niklas, 1994; Návar, 2010a) and the mean scalar exponent, according to model [21] and [22], is consequently $B$ = 2.53. Models [21] and [22] assume that the volume equation has an exponent of $D^{2.0}$. Návar (2010a) using the shape-dimensional analysis coupled with fractal geometry noted that d = 1.93 (0.066) and h = 0.917 (0.079) for 12 volume equations for temperate trees of northwestern Mexico. Therefore, an exponent value d ~ 1.9 (0.07) would be appropriate for these forests. That is, boles are neither two dimensional photosynthetic surfaces ($D^2$) nor three dimensional geometric solids ($D^2H$); hence, if d ~ 1.9, then $B$ = 2.43 in the Ketterings et al. (2001) or Chavé et al. (2005) semi-empirical models. This new slope value falls within the confidence bounds of the mean $B$-value found in Meta analysis studies (2.38 ±0.06).

The major finding of this brief review is that most current semi-empirical and theoretical studies assume a constant $B$-scalar exponent value. That is: $B_{\text{Návar}} \leq B_{\text{Chave}} = B_{\text{Ketterings}} \leq B_{\text{West}}$; 2.38 ≤ 2.53 = 2.59 ≤ 2.67. Further empirical and theoretical studies are required before the constant $B$-scalar exponent value finally emerges.

### 2.12 Implications of reduced non-destructive models of M assessment

Reduced non-destructive models that assume a constant $B$-scalar exponent easily calculates M for each individual tree as well as for any set of trees since it depends upon the $a$-scalar intercept value that is a function of the wood specific gravity value. The major implicit hypothesis of a reduced model such as the WEB or the Návar (2010b) equations would then be that trees add mass, volume, area or length at a rate per unit of diameter growth that is a function of the $a$-scalar intercept, which is a function of the $\rho_w$ values. Návar (2010b) found a positive relationship between $a$ and $\rho_w$, consistent with the explicitly statement described in the theoretical and semi-empirical models. If so, then trees with large $\rho_w$ figures would grow diametrically (as well as to any other dimension) at a small rate and vice versa, since D = $\sqrt[B]{\frac{M}{a}}$. A preliminary analysis of diameter increment and $\rho_w$ values for 15 tropical species fitted well with a negative linear relationship with the following equation: $\frac{\partial D}{\partial t} = 4.23 - 5.38\rho_w$ ; $r^2$=0.50; further empirically supporting the evidence that a reduced non-destructive, semi-empirical or the theoretical model that assumes a constant *B*-scalar exponent is also physiologically and metabolically correct.

The selection of a constant *B*-scalar exponent value in a reduced semi-empirical model has several consequences. Statistically, the *B* and *a* scalar coefficients are related with negative power or logarithmic equations (Zianis & Mencuccini, 2004; Pilli et al., 2006; Návar, 2009a,b). Hence, the *a*-scalar intercept would deviate from values reported in most allometric studies by assuming a different *B*-scalar exponent. For example, Table 1 reports mean (confidence bound) population values for the *a*-scalar intercept as: 0.14 (0.03). Therefore, when assuming a different scalar exponent values either the taper factors (C) or the basic specific gravity (ρ) for the entire tree would also change. Since ρ is assumed to be a fixed value for any tree, then the *a*-scalar intercept must have a fixed value as well that is only dependent upon the C coefficient.

The C values would be later more precisely and physically evaluated as long as new information and data analysis comes up. In the meantime, Návar (2010b) and Návar (2010d) have noted that the C empirically-estimated value when plotting $\rho_w$ vs. *a* varies between 0.2457 to 0.2687 for biomass equations reported for temperate North American and for tropical tree species, respectively. When assuming that *B* = 2.38, good tree M approximations are found for temperate and some tropical but not for dry land tree species. If further assuming that *B* = 2.67, tree M is overestimated for both temperate and for tropical forest communities. Whence, a C coefficient value should be further calculated with this later assumption by $C_{B=2.67} = (0.2457D^{2.38})/(D^{2.67})$. Again the C value is a function of D and it can go from 0.18 in trees with D = 5 cm to 0.076 in trees with D = 100 cm; following a power function of $C_{B=2.67} = 0.2457D^{(2.38-2.67)} = 0.2457D^{(-0.29)}$.

An independent technique to estimate the C coefficient figure was preliminarily proposed by Návar (2010b) by developing the shape-dimensional analysis as C = ($a_v$*$a_h$). Mean (standard deviation) $a_v$ values of 0.55 (0.0185) were found when fitting the statistical coefficients of the Schumacher and Hall (1933) volume equation to 12 temperate tree species of northern Mexico. By assuming a mean *a*-re-calculated scalar intercept value of 0.12 (Table 1) and the mean (standard deviation) of the taper values by solving for the $a_h$ values since they are hard to find at this time, the C coefficient would attain a range of 0.2104-0.2249 for 68% or 0.2037-0.2330 for 95% of the individual biomass equations, assuming the proportionality coefficient is normally distributed. The $C_{B=2.38}$ (0.2457±0.0152) for temperate North American tree species is found within this range. For tropical tree species (0.2687±0.1078), it appears to be slightly overestimated. On the other side, the $C_{B=2.67}$ (0.076-0.180) values are a tone with the C coefficient (0.11) proposed by Ketterings et al. (2001) but both are underestimated when contrasting them with C range values proposed by the shape-dimensional analysis. The C coefficient value proposed by Ketterings et al. (2001) is dependent upon $\rho_w$ since it was calculated as: C = $\rho_w/a$. From the shape-dimensional or the fractal analysis, C = $a/\rho_w$. New approaches on how to analyze biomass data will eventually elucidate the value of C and $a_h$. One way to go is to analyze backwards biomass data to solve for C or by $a_h$ when applying the empirical conventional allometric model [1]. For example; when fitting the WBE model, the C coefficient could be evaluated by: C = $M/(\rho_w D^{2.67})$ or when developing the semi-empirical model derived from the shape-dimensional analysis; $a_h = M/(a_v \times \rho_w D^d + H^{H^*})$ and then $C = (a_h \times a_v)$; or by evaluating C mathematically or iteratively until finding the right solution for each allometric biomass equation and then solving backwards for $a_h$. This research would eventually find the right semi-empirical, non-destructive methodology for the simple estimation of the scalar coefficients that facilitate tree M assessments. The slenderness $a_h$.parameter is related to the D/H relationship. However, further mathematical evidence is required to better calculate this parameter.

Independent preliminary C coefficient estimates approximated by $C = M/(\rho_w D^{2.38})$ or by $C = M/(\rho_w D^B)$ for several clusters of tree species are reported in Figure 7. The mean C coefficient values are 0.27 (0.09) and 0.25 (0.04) for the conventional and Návar (2010b) models and this preliminarily analysis demonstrate that these figures are also consistent with previously reported values for temperate (C = 0.2457±0.0152) and for tropical forests (C = 0.2687±0.1078). Statistical differences are noted between clusters of species, with C values smaller in Cedar/Larch, Pine, and Fir/Hemlock than in Maple/Hickory or Douglas Fir.

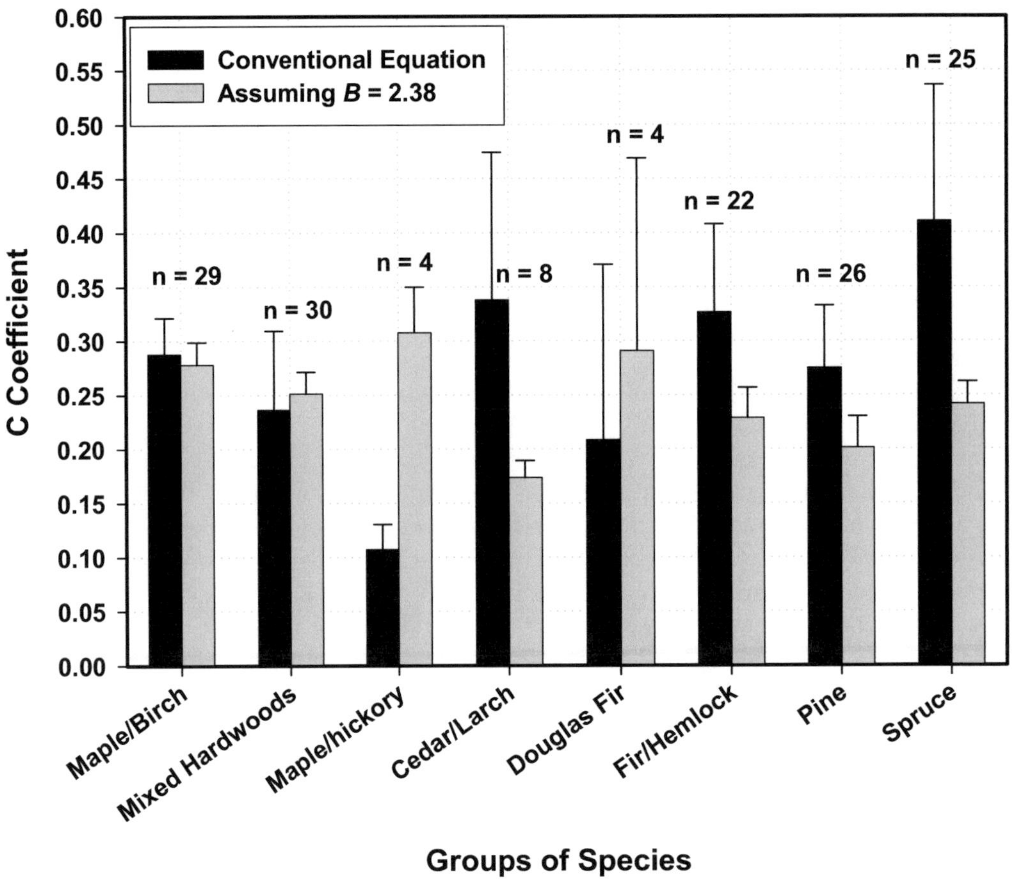


Fig. 7. C coefficient values of the conventional and Návar (2010b) reduced models for eight groups of species. Mean and confidence bounds (p = 0.05) are also depicted.

## 2.13 Calculating the entire tree specific gravity value

Several recent allometric studies include the wood specific gravity value as an exogenous variable (Brown, 1997; Chavé et al., 2005; Návar, 2009a). The theoretical WBE model calls for the specific gravity value for the entire tree (West et al., 1999). The Návar (2010b) reduced semi-empirical model requires the wood specific gravity value to escalate the *a*-scalar intercept value. The $\rho_w$ values are conventionally measured and reported figures can be found in recently-reported compilations. There remains the question to be solved whether $\rho_w = \rho$. Therefore, further information is required on easy ways to estimate $\rho$. The physical assessment of a weighted $\rho$ value for the entire tree is derived with model [23]:

$$\rho = \frac{M_b \rho_w + M_{br} \rho_{br} + M_l \rho_l}{M} \tag{23}$$

Where: $M_b\rho_w$ = bole mass x bole specific gravity; $M_{br}\rho_{br}$ = branch mass x branch specific gravity; $M_l\rho_l$ = leaf mass x leaf specific gravity; M = total aboveground biomass.

The mean $\rho_w$ value is difficult to assess, since it changes from the bark to the pith and from the bole base to the tip (Parolin, 2002). Miles and Smith (2009) compiled a comprehensive dataset containing standard specific gravity values for bole wood and bark for 156 North American tree species that can be preliminarily explored. Standard $\rho_w$ value is measured at DBH. Therefore, this as well as any other data source requires a correction factor since $\rho_w$ = f(H) (Silva-Arredondo and Návar-Cháidez, 2009). One approximation to solve for the change of $\rho_w$ with H is mathematically described in model [24]:

$$\begin{aligned} \rho_b &= \frac{M}{V} = \frac{Mc + Mh + Ms}{Vc + Vh + Vs} \therefore \\ Mc,h,s &= 0.7854 \int_{h=0}^{h=H} \left[ d = f(h) \cdot \rho_w = f(h,D) \right] \partial h \\ Vc,h,s &= 0.7854 \int_{h=0}^{h=H} \left[ d = f(h) \cdot \right] \partial h \end{aligned} \tag{24}$$

Where: M = mass, V = volume, c = bark, h = hardwood, s = softwood, h = relative tree height, H = total tree height, d = relative diameter, D = diameter.

Taper functions that relate bole diameter to relative bole tree height (d = f (h)) are available for most worldwide timber commercial tree species. Preliminary exploration of the bole wood and bark specific gravity data values reported by Miles and Smith (2009) show that they have similar specific gravity mean values (0.483 and 0.487) for all 156 North American tree species, although there are significant differences between these biomass components within each reported tree species. The scientific literature hardly reports leaf specific gravity values. However, leaf biomass accounts for by approximately 20% of the total aboveground biomass for 110 young trees of five species of northwestern USA (Delaney, 2007) and for 55 young trees of five pine species of Durango, Mexico (Návar et al., 2004). This ratio would probably steadily diminish with tree age and this relationship is also needed for most tree species.

Preliminary empirical evaluations of the entire tree basic specific gravity can be derived from the relationship between re-calculated *a* vs. ($a_v$ x $a_h$) or *a* vs. C with a slope describing $\rho$. Mathematically, $\rho = M/(C \times D^{2.38})$. These relationships are species specific and require sufficient allometric data to obtain coefficients with minimum variance. Site features may also influence this statistical function. Therefore, additional studies are needed to further advancing the knowledge on simple ways to estimate ρ and C with the goal of developing improved non-destructive methods of M assessment. In the meantime bole wood specific gravity is an estimator of the entire tree specific gravity.

## 3. Plot aboveground biomass assessments

### 3.1 Introduction

The application of allometric equations to forest inventory tree data is the standard methodology for the plot, stand M assessment since allometric equations that straightforward calculate plot aboveground biomass are hard to find. One equation for tropical dry forests of Jalisco, Mexico reported by Martínez-Yrizar et al. (1992) and a second one for semi-arid sub-tropical Tamaulipan thornscrub matorral of northeastern Mexico published by Návar et al. (2002b) were found in a brief scientific literature review. One major drawback of these equations is that they harvested all standing trees, including un-inventoried trees (d < 7.5 cm) in small plots (5 m x 5 m). Both equations use basal area as independent variable and the second one draw on also mean stand H and tree diversity, S, as an index of the stand wood specific gravity variation.

The conventional physics equation (model [19]) is an independent method for calculating tree and plot biomass. Commercial and research forest inventories conventionally report timber volume, V, at the plot scale. The entire tree specific gravity is at the present difficult to calculate with reported information, but an estimator can be used instead, the bole wood specific gravity, $\rho_w$, that is conventionally measured and reported in most wood technology studies. One shortcoming of this procedure is that branch volume is rarely integrated into bole volume estimates. On the other hand, Silva-Arredondo and Návar-Cháidez (2009) reported that the mean $\rho_w$ value for trees should be taken at 0.22H for temperate trees of northern México when guidelines for $\rho_w$ measurements recommend taking core samples at diameter at breast height, and therefore M estimates must use a correction factor to assess less biased M values.

A second independent approach to evaluate plot M multiplies plot volume times a biomass expansion factor, BEF. The BEF values previously calculated by the ratio of M/V are available for several tree species and for several forests (Gracia et al., 2004; Lehtonen et al., 2004; FAO, 2007; Návar-Cháidez, 2009; Silva-Arredondo and Návar, 2009). Whenever FEB, V and $\rho_w$ data are available, M estimation procedures described above can be used as contrasting methods since they are partially independent methods of plot M assessment. Brown (1997) coupled these variables to come up with plot M density with the following equation:

$$M = ER \cdot \rho_w \cdot BEF \qquad (25)$$

Where ER = stand timber volume ($m^3\ ha^{-1}$); $\rho_w$ = mean standard bole wood specific gravity for the tree species dominating the stand ($Mg\ m^{-3}$); and BEF = biomass expansion factor (dimensionless).

The BEF value of equation [25] is dimensionless and it only expands bole plot M to the entire aboveground tree biomass (boles, branches and leaves). Brown (1997) interpolated this equation for complex forests by weighting it for tree species or genera that constitute the forest.

The conventional BEF dimensional values reported in the recent scientific literature take the following form:

$$BEF = \frac{M(Mg)\,or\;(Mg\,ha^{-1})}{V(m^3)\,or\;(m^3\;ha^{-1})} = BEF\left[\frac{Mg}{m^3}\right] \tag{26}$$

Preliminary analysis of BEF values point to the local calibration by mean stand H, D, age, density, etc. to be recommended as methods of plot M assessment (Brown, 2002; Gracia, 2004; Lehtonen et al., 2005; Návar-Cháidez, 2009; Silva-Arredondo & Návar-Cháidez, 2009; Návar, 2010d). Therefore they are currently empirical factors of local use. The FAO (2007) compiled dimensional BEF figures for worldwide forests and they are shown in Table 2.

| | 1 m³ of timber volume is equivalent to: | |
|---|---|---|
| Region/Sub-region | Aboveground Biomass (Mg) | Total tree biomass (Mg) |
| Eastern Africa | 2.3 | 2.9 |
| North Africa | 2.1 | 2.7 |
| Western and Central Africa | 1.3 | 1.7 |
| Africa | 1.5 | 1.9 |
| Eastern Asia | 0.7 | 0.9 |
| Southern Asia | 1.4 | 1.8 |
| Western and Central Asia | 0.9 | 1.1 |
| Asia | 1.1 | 1.4 |
| Europe | 0.7 | 0.8 |
| Caribbean Countries | 2.0 | 2.6 |
| Central-America | 1.4 | 1.8 |
| North America | 1.0 | 1.1 |
| North and Central America | 1.0 | 1.2 |
| Australia New Zealand | 1.4 | 2.0 |
| South America | 1.1 | 1.5 |
| World | 1.0 | 1.3 |

Total tree biomass = boles, branches, foliage and roots.

Table 2. Biomass expansion factors to assess below and total standing stand aboveground biomass as a function of bole volume (Source: FAO, 2007)

Reported BEF values are practical for regional aboveground and total tree biomass calculations. For specific, local biomass projects, regional BEF factors must be applied whenever they are available since they can vary notoriously from place to place by changes in the forest structure (Brown, 2002).

Most studies that evaluate standard plot aboveground biomass use a single mathematical function, which is frequently the most popular, the one developed for worldwide studies, or the one locally developed, although there is a wide range of allometric equations available for regional and world-distributed forest communities (Brown et al., 1989; Brown, 1997; Chavé et al., 2001; 2003; 2005; Ketterings et al., 2001; Návar et al., 2004; Návar, 2009a,b). Therefore, when contrasting plot M estimates with different allometric equations or methods, they will always show variations.

### 3.2 Contrasting plot M assessments

Research studies that contrast plot M estimates by different allometric models or different methods of estimation are scarce in the scientific literature. Figure 2 exemplifies the large variation expected when evaluating tree M for tropical rainforest, tropical dry forests, as well as for the hybrid IAN 710 *Hevea brasiliensis* trees. Figure 4 also shows large differences in M evaluations for similar tree species. Therefore, the selection of an allometric model is the most important source of uncertainty when assessing plot M (Chavé et al., 2004; Návar et al., 2010).

Several other uncertainty sources can be expected when computing tree and consequently plot M. The allometric equation has an intrinsic error given by the standard deviation of the tree M estimate that is related to the number, diversity and diameter structure of sampled trees for biomass allometry. The second, and probably the most important, arises when an off-site equation is applied to forests with different tree diversity and diameter structure. This error source has not been evaluated since allometric equations are hardly validated and therefore it is preliminary assessed when several allometric equations estimate plot biomass using the same forest inventory dataset (Chavé et al., 2004). Three other types of uncertainties related to the forest inventory scheme are: the size and shape of sampling plots, the spatial distribution of sampling units in the forest, and tree measurements within sampling plots.

Large variations in tree and consequently in plot M evaluations make the biomass stock of most forest stands to remain poorly understood. For example, for mid and high latitude forests, Fang et al. (2006) reviewed the literature and noted that inventory-based forest M stocks documented for major countries fall within a narrow range of 72–112 Mg $ha^{-1}$ with an overall area-weighted mean of 87.2 Mg $ha^{-1}$. These estimates are 0.40 to 0.71 times smaller than those (122–216 Mg $ha^{-1}$) computed in previous analysis.

For structurally-complex tropical forests, the evaluation of plot M appears to have larger variability (Fearnside and Laurance, 2003; Houghton, 2005; Saatchi et al., 2007). Chavé et al. (2003) tested four different allometric equations (Chavé et al., 2001; Chambers et al., 2001; Brown et al., 1989 (1); Brown et al., 1989 (2)) for Panamanian tropical forests and mentioned that all four equations yielded comparable but statistically different plot M estimates. The variation among equations was 26% of the mean estimate. Houghton et al. (2001) tested seven different plot M estimates for the Brazilian Amazon forests and calculated a mean deviation of 20% but individual observations deviated 45% from the mean estimate. Araujo et al. (1999) harvested and weighed all standing tree biomass in a 0.2 ha area plot of the Brazilian Amazon forest. Of 14 different biomass equations applied to this dataset, 12 biased notoriously and only two provided suitable plot M assessments, within ± 0.6% of the weighted field biomass.

### 3.3 An example of the application of tree allometry for the plot M assessment of Mexican tropical forests

Návar et al. (2010) evaluated plot M by fitting nine different allometric equations for eight different Mexican tropical forests. Fitted functions are reported in Table 3.

The conventional physics equation of volume times the wood specific gravity was also fitted. Results showed that several allometric equations predicted significantly different mean stand M estimates for all eight data sources (Figure 7). The mean deviation between allometric equations was 10.7 Mg $ha^{-1}$ (62%) and uncertainties are a function of the forest aboveground biomass density ($Sxe = 3.92M^{0.71}$; $r^2 = 0.92$;). Therefore the allometric equations consistently yielded larger mean standard errors for moist and wet than for dry tropical forests.

| Researcher | Tropical Forest | Equation |
|---|---|---|
| Brown et al. (1989) | Wet | $M=e(-2.409+0.952(\rho_w D^2H)$ |
| Brown (1997) | Dry | $M=34.47-8.0671D+0.6589D^2)$ |
| Brown (1997) | Moist | $M=e(-2.134+2.53*\ln(D))$ |
| Martínez-Yrizar et al. (1992) | Dry | $M=10**(-0.5352+Log_{10}(BA))$ |
| Návar et al. (2010) | Dry | $M=0.08479(\rho_w{}^{0.55255}D^{2.2435}H^{0.4773})$ |
| Návar et al. (2010) | Dry | $M=e(-2.409+0.952*Ln(D^2H))$ |
| Návar (2010a) | All | $M=(38.36*B^{-6.9045})D^{(B=d+hB^*)}$ |
| Návar (2010b) | All | $M=(-0.0094+0.2687\rho_w)D^{2.38}$ |
| Chavé et al., (2005) | Dry | $M=0.112*(\rho_w D^2H)^{0.916}$ |
| | Dry | $M=\rho_w*e(-0.667+1.784Ln(D)+0.207ln(D)^2-0.028Ln(D)^3$ |
| Chavé et al., (2005) | Moist | $M=0.0509*(\rho_w D^2H)$ |
| | Moist | $M=\rho_w*e(-1.49+2.148Ln(D)+0.207ln(D)^2-0.028Ln(D)^3$ |
| Chavé et al., (2005) | Wet | $M=0.0776(\rho_w D^2H)^{0.94}$ |
| | Wet | $M=\rho_w*e(-1.24+1.98Ln(D)+0.207ln(D)^2-0.028Ln(D)^3$ |

Where: The non-destructive method proposed by Návar (2010a) uses the H-D and V=H,D (i.e., $H = a_hD^{B^*}$ and $V = a_vD^dH^h$) equations and an empirical equation to estimate $a$ = $38.36B^{-0.6.9045}$; D = diameter at breast height, BA = basal area; $\rho_w$ =wood specific gravity.

Table 3. Allometric equations employed in aboveground biomass estimation for eight tropical forest communities of Mexico.

Total aboveground biomass deviations for all tropical forests inventoried between either a) allometric equations or b) allometric and the physics equation were smallest when using the non-destructive model proposed by Návar (2010a). The equations of Brown (1997) for tropical dry forests and the equation of Návar (2010a) yielded consistent plot M estimates across allometric equations and methods. Deviations as large as 2.4 orders of magnitude were found; i.e. see for example stand M assessments by the Brown (1997) and the physics equation for moist forests or the non-destructive method II and the local equation developed by Návar (2009) for tropical dry forests.

### 3.4 Future directions in stand M assessments

The large deviations due to the application of allometric models or methods developed off site reveal that probabilistic plot M estimates are highly likely to be skewed by more than one and sometimes up to two orders of magnitude. In the absence of harvested trees in sampling stands for the development of local plot M allometry, the choice of tree biomass models must focus on those that consistently result in similar plot M estimates and with the least deviation to the physics equation. Whenever it is possible, a mean estimate across allometric equations that do not notoriously deviate between them would probabilistically deliver a better plot M assessment. However further research is required on this issue for several complex forests before it is recommended to improve precision of plot M assessments.

### 3.5 The need for the development of plot allometry

In future plot allometry studies, research must center on harvesting all inventory trees from plots under, for example, by shifting cultivation, the opening of roads, or other forest disturbances. Weighting all biomass, collecting samples for fresh and dry weights and

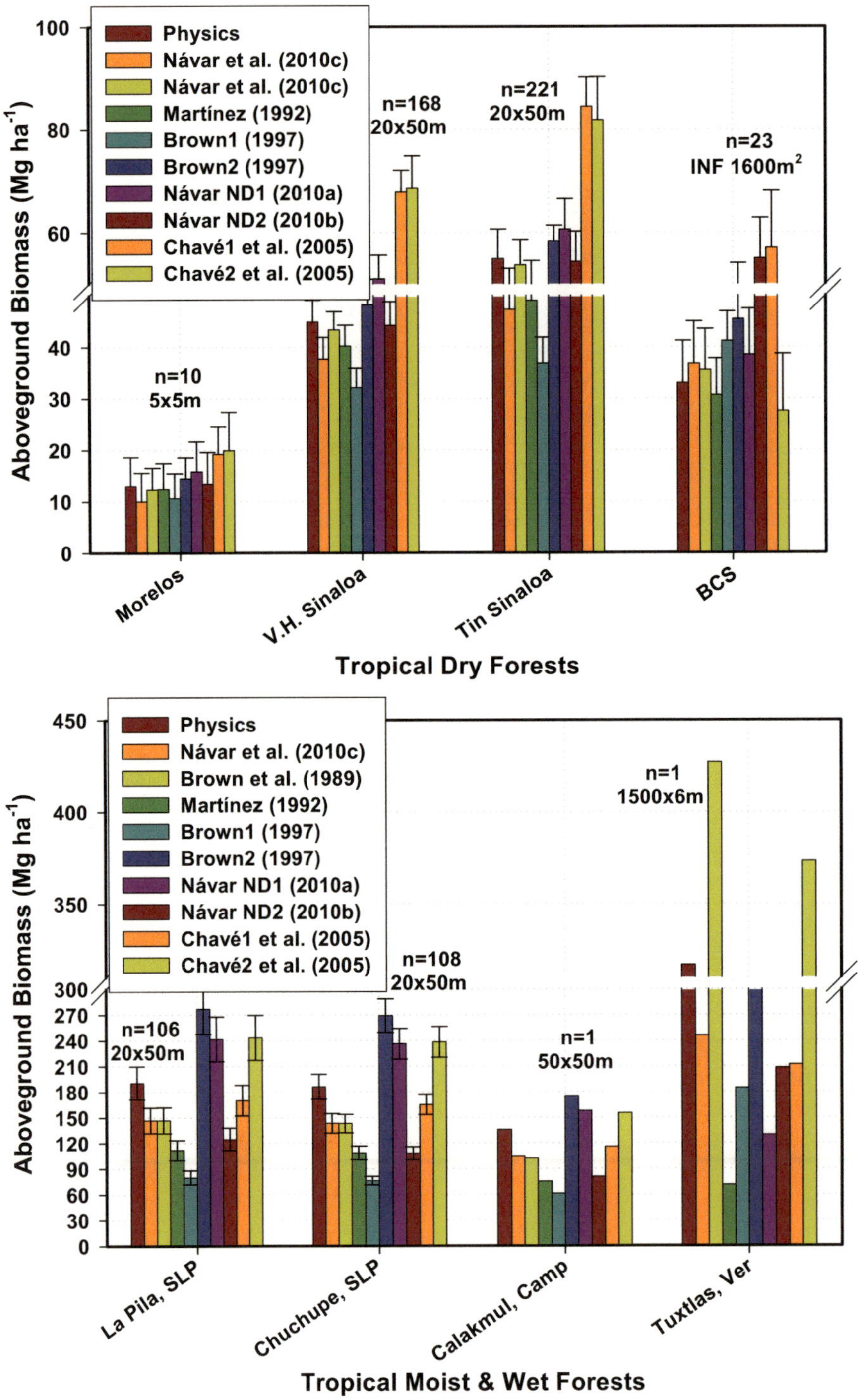


Fig. 8. Mean and confidence interval aboveground biomass estimates for two tropical forests calculated by different allometric equations for eight tropical forest communities of Mexico.

calculating all stand M. The application of local or worldwide-developed tree allometry must be the first step to calculate stand M. In case these techniques do not provide a good fit, new tree and plot M equations should be developed that can be escalated up and down from tree levels to plot or stand scales.

Other alternatives that expand local allometry to plot scales could be the combination of restrictive in the number of harvested trees (Zianis & Mencuccini, 2004) and non-destructive (West et al., 1997; Návar, 2010a; 2010b) methods with local tree or plot allometry. Whenever local allometric equations are not available, coupling available empirical equations, restrictive models and semi-empirical non-destructive methods for inventoried datasets and choosing an average of those that consistently produce similar plot M evaluations could improve stand M precision, according to the Central Limit Theorem. However there are a few of such studies reported in the scientific literature that discus this issue.

## 4 Biomass assessments at regional scales

### 4.1 Introduction

Tree or plot M assessments are extrapolated over larger spatial scales to evaluate forest, regional or national aboveground biomass, AGB. Houghton et al. (2001) classify techniques available in the interpolation procedure as: a) field measurements, b) environmental gradients and c) remote sensing techniques. Field measurements are common approaches and they are mathematically just the multiplication of the mean plot M estimate times the area of each forest type (Schroeder et al., 1997; Houghton et al., 1992; Brown, 2002). Commercial and research forest inventories or a combination of both field data collection sources provide information for the AGB estimation for each forest class. This approach is time-consuming, labor intensive, expensive, and difficult to implement, especially in remote areas and it cannot provide the spatial distribution of AGB biomass for large areas because the error involved increases with the forest area. The major assumption is that a good sampling scheme represents the forest class under research. The error can be evaluated by multiplying the standard deviation, the standard deviation as a percentage of the mean, percent error, or the confidence interval over the forest area. The simple field method can increase precision by augmenting the spatial resolution of each forest type combined with the application of improved sampling schemes. AGB assessments augment precision when this method is applied for compact forest classes such as old-growth forests and forests plantations.

The environmental gradient method should be the recommended technique when there are systematic plot M changes over large forest areas. So far, annual rainfall has proved to be the best predictor variable for Amazonian rainforests (Saatchi et al., 2007) as well as for Mexican tropical forests (Návar et al., 2010c). In general, non-linear models account for by between 50 to 60% of the total stand M variation and they are constrained to attain a final steady constant M value indicating that highest M figures do not have further increments with additional annual rainfall. However, for seasonal or dry forests the relationship is almost linear stressing the importance of the moisture available in places with ample light and radiation (Malhi et al., 2006). In this case, the integration of the non linear equation relating stand M to annual rainfall multiplied by the area evaluates AGB. This relationship can de discretized at different spatial resolutions by calculating forest areas with similar annual rainfall that are, in general, areas with the same altitude above sea level. The error involved in this relationship is associated with the standard error of this relationship which is also multiplied by the same area.

Remote sensing techniques are currently indirect methods of AGB assessments. Remote sensing based methods for AGB assessments are classified as: a) fine spatial resolution (aerial photographs, IKONOS), b) medium spatial resolution data (Landsat, Spot, TM/ETM+), c) coarse spatial resolution data (AVHRR, IRS-IC, WiFS), d) methods based on radar data (Radar, Lidar) (Lu, 2006). Tree M can be better computed when using fine spatial resolution methods as well as radar data. Stand M is better evaluated when the image spatial resolution is from medium to coarse. Remote sensing is also the standard methodology used to classify forests according to vegetation types, and/or cover classes (Houghton et al., 2000; 2001; Saatchi et al., 2007) and takes advantage of the high correlations between spectral signatures, vegetation parameters, transformed images, and image textures to evaluate AGB using field measurements or environmental methods (Lu, 2006). Interpolation techniques are classified as: multiple regression analysis, nonparametric k-nearest neighbor technique, neural networks, or through indirect relationships between remotely sensed forest attributes and biomass (Popescu et al., 2003; Zheng et al., 2004).
Satellite data may eventually be developed to straightforward estimate AGB, but neither optical nor radar data have yielded consistent results in forests with moderate to high biomass (Nelson et al., 2000). Optical and radar data saturate with canopy closure or at relatively low levels of biomass, respectively (Rignot et al. 1997; Nelson et al. 2000). A promising new approach is the use of lidar, which yields a measure of tree height that is related to biomass (Means et al. 1999). Recent 3D models with lidar suggest that the data are highly correlated with aboveground biomass in coniferous forests of northwestern USA (Lefsky et al., 1999a; Means et al., 1999) as well as in deciduous forests of eastern USA (Lefsky et al. 1999b). A NASA satellite equipped with a lidar designed to measure tree heights, biomass, and topography, the Vegetation Canopy Lidar (VCL), is currently taking information and a new tree height world wide map is already on the NASA homepage.

### 4.2 Uncertainties of AGB evaluations

The variety of remote sensed data, spatial resolutions, tree and forest attributes, and interpolation techniques make AGB assessments variable. Therefore, the largest AGB uncertainty over large regions results from the extrapolation technique used. Houghton et al. (2001) evaluated seven different M interpolation methods (three field measurements, two methods based on environmental gradients and two methods using remote sensing techniques) with different spatial resolutions for the Brazilian Amazon forests. As a result, AGB estimates deviate by more than one order of magnitude, from 80 to 190 Pg and models also differed on the spatial AGB distribution (Houghton et al., 2001). Although most research has been conducted for tropical forests, Houghton et al. (2001; 2003; 2005); Jenkins et al. (2003); and Pacala et al. (2001) pointed out that forest biomass for the mid-and high latitudes in the Northern Hemisphere is also uncertain due to error in the estimation of tree and plot biomass and its increment. For instance, biomass estimates of Russian forests varied from 112.0 Pg to 140.4 Pg by different authors (Alexeyev et al., 1995; Isaev et al., 1995) although the same forest data sources but different methods were used.

### 4.3 An example of AGB assessment for Mexican tropical forests and contrasting results

For Mexican tropical forests, Návar et al. (2010c) contrasted two field measurements (red and blue bars), one environmental gradient (green bar), and one Forest Inventory Analysis

(FIA) to assess AGB. In addition, AGB estimates reported by De Jong et al. (2008); Houghton et al. (1999) and De Fries et al. (2002); Brown (1997) and Achard et al. (2004); and IPCC (2006) integrated a contrast analysis. Návar (2010c) calculated that AGB assessments vary between 1.54 with the FIA to 3.00 Pg with the improved spatial resolution analysis. Other evaluations ranged from 3.84 to 4.34 Pg (Figure 9), which are larger than AGB assessments conducted by the author of this report.

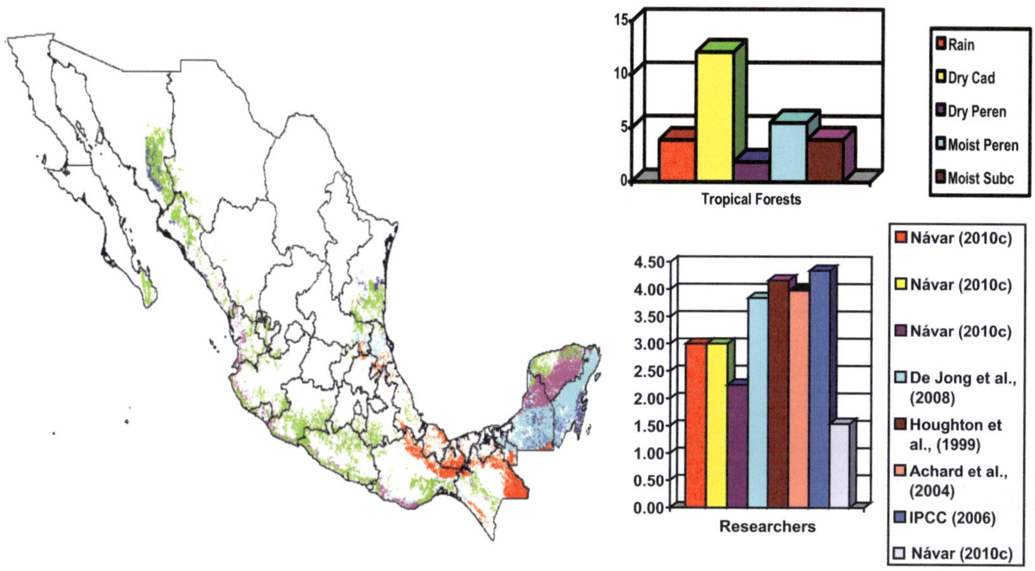


Fig. 9. Mean aboveground biomass estimates for Mexican tropical forests calculated by different methods of interpolation (Red = rain, green = dry deciduous, purple = moist deciduous, and blue = moist evergreen forests). Návar-Cháidez (2009) reported biomass expansion factors, BEF, for dry tropical forests of Sinaloa, Mexico as 1.46 (0.022) that are quite consistent with the BEF values reported by FAO (2007) for Latin American tropical forests. Using the total standing volume value calculated in the Mexican Forest Inventory of 2004-2006, AGB biomass expands to 1.54 (0.023) Pg.

The FIA estimate was 2.8 orders of magnitude smaller than the AGB estimate made by IPCC (2006). Assuming all these eight AGB statistics yield a mean of 3.26 Pg, deviations as large as 1.72 Pg (52.76%) are found between methods of M interpolation at the regional scale with a mean absolute deviation of 0.82 Pg (25% of the mean value).

#### 4.4 Future directions in regional AGB evaluations

The application of a single interpolation methodology from tree or plot M to larger spatial scales is also highly likely to skew regional AGB figures. Therefore, there is an urgent need for combining available methods of AGB computation to understand the magnitude and sources of variation. Future studies must focus on using at least a second independent interpolation approach and an average estimate and its deviation should be reported. However, I am tempted to recommend the coupling of all three current methodologies of AGB calculation. These are: field techniques, collected in most forest inventory studies; develop relationships between environmental variables (annual precipitation, evapotranspiration, water and energy balance, Bowen ratio, etc.) and plot M stock estimates and fluxes that are calculated with proved allometric studies; with forests classified by field and remote sensing techniques in several classes and gradients within classes. Although the combination of these techniques requires high expenditure of resources, a diversity of professional skills, and improved current technologies, more precise AGB approximations would be expected that may eventually reduce costs of regional biomass assessments. A single, independent approach could be to develop tree data (diameter and canopy height) with LIDAR techniques; use empirical, locally-derived volume and canopy height equations ($v=f(D,H)$; $H=f(D)$) to be fitted to the semi-empirical shape-dimensional, nondestructive model to evaluate tree M at the spatial scale of interest. This is a matter of future studies.

## 5. Conclusions

Many tree species and worldwide forests do not have updated local allometry. They call for the development and application of local allometry. However, there are a great number of allometric equations reported for worldwide species with a major emphasis on temperate North American and European tree species. Available equations can be classified by the way equation parameters are estimated as empirical, theoretical and semi-empirical non destructive methods of tree M assessment. Empirical models commonly calculate statistical coefficients by least square techniques in linear, log-linear, non-linear, multiple linear and non-linear, seemingly unrelated linear and non-linear and exponential mathematical formats where diameter at breast height or basal diameter, top height, canopy area and wood specific gravity are independent variables that individually or in combination explain tree M of harvested trees. Conventional mathematical allometric models have intrinsic variations larger than 20% of the tree M but new empirical equations that contain H and $\rho_w$ as independent variables, in addition to the conventional normal diameter, are improving tree M evaluations since the intrinsic error has been reduced to 16%. Theoretical and semi-empirical non-destructive methods are in the early stages of development and require further testing and refinement although they can be preliminarily recommended as non-destructive approaches of tree and stand M computation. In spite of the wealth on biomass allometry for several trees and forests, conventional plot M evaluations exhibit variations as large as two orders of magnitude when contrasting on and off-site equations. One potential procedure to reduce uncertainty is combining the conventional empirical, semi-empirical non-destructive, and restrictive methods to improve precision when computing tree and plot M for forests that do not convey local biomass allometry. However, research on understanding variations by coupling these methods are lacking elsewhere. There is a variety of interpolation techniques of tree or plot M to forests, regional and national scales but they display variations close to three orders of magnitude when assessing AGB stocks. Coupling tree allometry with FIA evaluations, environmental gradient analysis and remote sensing techniques may reduce this variation in future AGB studies. Modern remote tree data collection techniques (LIDAR) combined with empirical, locally-derived functions to estimate timber volume as a function of diameter and canopy height and canopy height as a function of diameter plugged into the non-destructive semi-empirical, shape-dimensional analysis model could improve AGB evaluations at the spatial scale of interest.

## 6. Acknowledgements

The senior author wishes to thank all anonymous reviewers for the improvement of the readability and technical content of this report.

## 7. References

Achard, F., Eva, H.D., Mayaux, P., Stibig, H.J. and Belward, A. (2004). Improved estimates of net carbon emissions from land cover change in the tropics for the 1990s Glob. Biogeochemical Cycles 18 GB2008 doi:10.1029/2003GB002142.

Alexeyev, V., Birdsey, R., Stakanov, V. and Korotkov, I. (1995). Carbon in vegetation of Russian forests: Methods to estimate storage and geographical distribution, Water Air Soil Pollution 82: 271- 282.

Araujo, T.M., Highughi, N., Carvalho, J.A. (1999). Comparison of formulae for biomass content determination in a tropical rain forest in the state of Para, Brazil. Forest Ecology and Management 117: 43–52.

Beskerville, G.L. (1965). Estimation of dry weight of tree components and total standing crop in conifer stands. Ecology 46: 867–869.

Beskerville, G.L. (1972). Use of logarithmic regression in the estimation of plant biomass. Canadian Journal Forest Research 2: 49–53.

Brandeis, T.J., Delaney, M., Parresol, B., Royer, L. (2006). Development of equations for predicting Puerto Rican subtropical dry forest biomass and volume. Forest Ecology and Management 233: 133–142.

Brown, S. (1997). Estimating biomass and biomass change of tropical forests. Forest Resources Assessment Publication. Forestry Papers 134. FAO, Rome, 55 pp.

Brown, S., Gillespie, A.J., Lugo, A.E. (1989). Biomass estimation methods for tropical forests with applications to forest inventory data. Forest Science 35, 881–902.

Brown, S. (2002). Measuring carbon in forests: current status and future challenges. Environmental Pöllution 116: 363-372.

Burrows, W.H., Hoffman, M.B., Compton, J.F., Back, P.V., and Tait, L.J. (2000). Allometric relationships and community biomass estimates for some dominant eucalypts in Central Queensland woodlands. Australian Journal of Botany 48: 707-714.

Cairns, M.A., Dirzo, R., and Zadroga, F. (1995). Forests of Mexico: a diminishing resource? Journal of Forestry 93: 21-24.

Cairns M A, Brown S, Helmer E H and Baumgardner G A. (1997). Root biomass allocation in the world's upland forests. Oecologia 111 1-11

Cairns, MA., P.K. Haggerty, R. Alvarez, BHJ De Jong, and I. Olmsted. (2000). Tropical Mexico's recent land use and change: a region´s contribution to the global carbon cycle. Ecological Applications 10: 1426-1441.

Cairns, M.A., Olmsted, I., Granados, J., Argaez, J. (2003). Composition and aboveground tree biomass of a dry semi-evergreen forest on Mexico's Yucatan Peninsula. Forest Ecology and Management 186: 125-132.

Castellanos, J., M. Maass, J. Kummerow. (1991). Root biomass of a deciduous tropical forest in Mexico. Plant and Soil 131: 225-228.

Castellanos, J., VJ. Jaramillo, RL. Sanford, and J.B. Kaufmann. (2001). Slash-and-burn effects on fine root biomass and productivity in a tropical dry forest ecosystem in Mexico. Forest Ecology and Management 148: 41-50.

Chambers, J.Q., dos Santos, J., Ribeiro, R.J., Higuchi, N. (2001). Tree damage, allometric relationships, and above-ground net primary production in central Amazon forest. Forest Ecology and Management 152: 73–84.

Cannell, M.G.R. (1984). Woody biomass of forest stands. Forest Ecology and Management 8: 299–312.

Chavé, J., Riera, B., Dubois, M.A. (2001). Estimation of biomass in a neotropical forest of French Guiana: spatial and temporal variability. Journal of Tropical Ecology 17: 79–96.

Chavé, J., Condit, R., Lao, S., Caspersen, J.P., Foster, R.B., Hubbell, S.P. (2003). Spatial and temporal variation in biomass of a tropical forest: results from a large census plot in Panama. Journal of Ecology 91: 240–252.

Chave, J., Andalo, C., Brown, S., Cairns, M.A., Chambers, J.Q., Eamus, D., Fölster, H., Fromard, F., Higuchi, N., Kira, T., Lescure, J.-P., Nelson, B.W., Ogawa, H., Puig, H., Riera, B., Yamakura, T. (2005). Tree allometry and improved estimation of carbon stocks and balance in tropical forests. Oecologia 145 (1): 87–99.

Chavé, J., R. Condit, S. Aguilar, A. Hernández, S. Lao, and R. Perez. (2006). Error propagation and scaling for tropical foret biomass estimates. Phil. Trans. R. Soc. London B. 359: 409-420.

Chavé, J., H C Muller-Landau, T R Baker, T A Easdale, H Steege and C O Webb. (2006). Regional and phylogenetic variation of wood density across 2456 neotropical tree species. Ecological Applications 16: 2356-2367.

Chojnaki, Colorado G.J. (2001). Ecuaciones de biomasa aérea para los árboles de los bosques secundarios del área de influencia de la Central hidroeléctrica Porce II. Trabajo de grado para optar por el título de Ingeniero Forestal, Universidad Nacional de Colombia, Medellín, Colombia.

De Jong, B.H.J., Cairns, M.A., Haggerty, P.K., Ramirez-Marcial, N., Ochoa-Gaona, S., Mendoza-Vega, J., Gonzalez-Espinosa, M., and March-Mifsut, I. (1999). Land-use change and carbon flux between 1970's and 1990's in central highlands of Chiapas, Mexico. Environmental Management 23: 373-385.

De Jong, B.H.J., Ochoa-Gaona, S., Castillo-Santiago, M.A., Ramirez-Marcial, N., and Cairns, M.A. (2000). Carbon flux and patterns of land-use/land-cover change in the Selva Lacandona, Mexico. Ambio 29: 504-511.

De Jong, B.H.J., Iglesias-Gutiérrez, L., and Alanís de la Rosa, J.A. (2008). Advances of Mexico in preparing for REDD. UNFCC Workshop. Tokyo, Japan.

DeFries R S, Houghton R A, Hansen M C, Field C B, Skole D and Townshend J. (2002). Carbon emissions from tropical deforestation and regrowth based on satellite observations for the 1980s and 1990s Proc. Natl Acad. Sci. USA 99 14256–61

Dirzo, R. and Garcia, C. (1991). Rates of deforestation in Los Tuxtlas, a netropical area in southeast Mexico. Conservation Biology 6: 84-90.

Enquist, B.J., Brown, J.H., West, G.B. (1998). Allometric scaling of plant energetics and population density. Nature 395: 163–165.

Enquist, B.J. (2002). Universal scaling in tree and vascular plant allometry: toward a general quantitative theory linking plant form and function from cells to ecosystems. Tree Physiology 22: 1045–1064.

Enquist, B.J., Niklas, K.J. (2002). Global allocation rules for patterns of biomass partitioning across seed plants. Science 295, 1517–1520.

FAO. (2007). Food and Agricultura Organization. State of the World's Forests. Rome, Italy.

Fearnside P.M., Laurance W.F. (2003). Comment on 'determination of deforestation rates of the world's humid tropical forests. Science 299: 1015a.

Feldpausch, T.R., McDonald, A.J., Passos, C.A.M., Lehmann, J., Riha, S.J. (2006). Biomass, harvestable area, and forest structure estimated from commercial forest inventories and remote sensed imagery in Southern Amazonia. Forest Ecology and Management 233: 121–132.

Fehrmann, L. and Kleinn, C. (2006). General considerations about the use of allometric equations for biomass estimation on the example of Norway spruce in Central Europe. Forest Ecology and Management 236: 412:421.

Flombaum, P., and Sala, O.E. (2007). A non-destrucvtive and rapad method to estimate biomass and aboveground net primary production in arid environments. Journal of Arid Environments 69: 352-358.

Gibbs, H.K., S. Brown, J.O. Niles, and J.A. Foley. (2007). Monitoring and estimating tropical forests carbon stocks: making REDD a reality. Environemntal Research Letters doi 10.1088/1748-9326/2/4/045023.

Gower, S.T., Kucharik, C.J., Norman, J.M. (1999). Direct and indirect estimation of leaf area index, fAPAR, and net primary production of terrestrial ecosystems. Remote Sens. Env. 70, 29–51.

Gracia, C., Vayreda, J., Sabaté, S., Ibáñez, J. (2004). Main components of the aboveground biomass expansion factors. CREAF, Centre de Recerca Ecologica i Aplicacions Forestals. University of Barcelona, Spain.

Houghton R A. (1999). The annual net flux of carbon to the atmosphere from changes in land use 1850–1990. Tellus B 51: 298–313.

Houghton, R.A., Hackler, J.L. and Lawrence, K.T. (1999). The US carbon budget: contributions from land-use change. Science 285: 574–578.

Houghton R A. Lawrence, K.T., Hackler, J.L., and Brown, S. (2001). The spatial distribution of forest biomass in the Brazilian Amazon: A comparison of estimates. Global Change Biology 7: 731-746.

Houghton R.A. (2005). Aboveground biomass and the global carbon balance. Global Change Biology 11 945–958.

Houghton R.A. (2005). Tropical deforestation as a source of greenhouse gas emissions Tropical Deforestation and Climate Change Ed. Mutinho and Schwartzman (Belem: IPAM).

Hughes, R.F., Kauffman, J.B., Jaramillo, V.J. (1999). Biomass, carbon, and nutrient dynamic of secondary forests in a humid tropical region of Mexico. Ecology 80: 1892–1907.

IPCC. (2006). IPCC Guidelines for National Greenhouse Gas Inventories. Prepared by the National Greenhouse Gas Inventories Programme ed H S Eggleston, L Buendia, K Miwa, T Ngara and K Tanabe (Japan: Institute For Global Environmental Strategies).

Isaev, A., Korovin, G., Zamolodchikov, D., Utkin, A. and Pryaznikov. A. (1995). Carbon stock and deposition in phytomass of the Russian forests, Water Air Soil Pollution 82: 247- 256.

Jaramillo, V.J., Kaufmann, J.B., Rentaría-Rodríguez, L., Cummings, D.L., Ellingson, L.J. (2003). Biomass, carbon and nitrogen pools in Mexican tropical forest landscapes. Ecosystems 6: 609-629.

Jenkins, J.C., Chojnacky, D.C., Heath, L.S., Birdsey, R.A. (2003). National-scale biomass estimators for United States trees species. Forest Science 49, 12–35.

Ketterings, Q.M., Noordwijk, C.M.Y., Ambagau, R., Palm, C.A. (2001). Reducing uncertainty in the use of allometric biomass equations for predicting above-ground tree biomass in mixed secondary forests. Forest Ecology and Management 146: 199–209.

King, D.A. (1990). The adaptative significance of tree height. The American Naturalist 135: 809-828.

Lefski, M.A., Cohen, W.B., Acker, S.A., Parker, G.G. and Shugart, H.H. (1999a). LIDAR remote sensing of the canopy structure and biophysical properties of Douglas fir western hemlock forests. Remote Sensing of the Environment 70: 339-361.

Lefski, M.A., Harding, D., Cohen, W.B., Parker, G.G. and Shugart, H.H. (1999b). Surface lidar remote sensing of basal area and biomass in deciduous forest of eastern Maryland, USA. Remote Sensing of the Environment 67: 83–98.

Lehtonen, A., Makipaa, R., Mukkonnen, P. (2004). Biomass expansion factors. Finnish Forest Research Institute. Helsinki, Finland. P29.

Litton, C.M., Sandquist, D.R., Cordell, S. (2006). Effect of non-native grass invasion on aboveground carbon pools and tree population structure in a tropical dry forest of Hawaii. Forest Ecology and Management 231: 105–113.

Litton, C.M., Kauffman, J.B. (2008). Allometric models for predicting aboveground biomass in two widespread woody plants in Hawaii. Biotropica 40: 313–320.

Lu, D. (2006). The potential and challenge of remote-sensing based biomass estimation. International Journal of Remote Sensing 27(7): 1297-1328.

Mandelbrot, B. (1983). Fractal Geometry of Nature. Freeman, New York.

Malhi Y, Wood D, Baker TR. (2006). The regional variation of aboveground live biomass in old-growth Amazonian forests. Global Change Biology 12: 1107–1138.

Martínez-Yrızar, A. M. Maass, L.A. Pérez-Jiménez, Sarukhán, J. (1992). Net primary productivity of a tropical deciduous forest ecosystem in western Mexico. Journal of Tropical Ecology 12: 169–175.

Masera, OR., Bellon, M.R., and Segura, G. (1995). Forest management options for sequestering carbon in Mexico. Biomass & Bioenergy 8: 357-367.

Means, J.E., Acker, S.A., Harding, D.J., Blair, J.B., Lefski, M.A., Cohen, W.B., Harmon, M.E. and Mckee, W.A. (1999). Use of large-footprint scanning airborne lidar to estimate forest stand characteristics in the western Cascades of Oregon. Remote Sensing of Environment 67: 298–308.

Miles, P.D., and Smith, W.B. (2009). Specific gravity and other properties of wood and bark for 156 tree species found in North America. USDA FS. Northern Research Station. Research Note NRS-38. Delware, OH. USA.

McMahon, T.A. (1973). Size and shape in biology. Science 179: 1201-1204.

Mohren, F. and Klein Goldewijkt, K. (1994). CO2 fix model. Institute of Forestry and Nature Research. Wageningen, Neetherlands.

Monroy C.R. and Návar, J. (2004). Ecuaciones de aditividad para estimar componentes de biomasa de Hevea brasiliensis Muell Arg. En Veracruz, México. Madera y Bosques 10: 29–43.

Návar, J., Nájera, A., y Jurado, E. (2002a). Biomasa estimation equations in the Tamaulipan thornscrub of northeastern Mexico. Journal of Arid Environments 52: 167-179.

Návar, J., Mendez, E., and Dale, V. (2002b). Estimating stand biomass in the Tamaulipan thornscrub of northeastern Mexico. Annals of Forest Sciences: 59: 813-821.

Návar, J., Gonzalez, N., Maldonado, D., Graciano, J., Dale, V., and Parresol, B. (2004a). Biomass equations for pine species of forest plantations of Durango, Mexico. Madera y Bosques 10(2): 17-28.

Návar, J., Mendez, E., Graciano, J., Dale, V., and Parresol, B. (2004b). Biomass equations for shrub species of Tamaulipan thornscrub of northeastern Mexico. Journal of Arid Environments 59(4): 657-674.

Návar, J. (2009a). Allometric equations for tree species and carbon stocks for forests of northwestern Mexico. Forest Ecology and Management 257: 427-434. doi:10.1016/j.foreco.2008.09.028

Návar, J. (2009b). Biomass component equations for Latin American species and groups of species. Annals of Forest Science 66: 208-216.

Návar, J. (2010a). A non-destructive approach for tree biomasa estimation. On Review in Forest Ecology and Management.

Návar, J. (2010b). A non-destructive, semi-empirical approach for tree biomass estimation. Submitted to Forestry. An International Forestry Review.

Návar, J., Pérez, G. and González, M. (2010). Stand biomass estimation in tropical forests of Mexico. On Review in Forest Ecology and Management.

Návar-Cháidez, J.J. (2009). Allometric equations and expansion factors for tropical dry trees of eastern Sinaloa, Mexico. Tropical and Subtropical Agroecosystems 10: 45-52.

Návar-Cháidez, J.J. (2010c). Biomass allometry for tree species of northwestern Mexico. Submitted to Tropical and Subtropical Agroecosystems.

Nelson, R.F., Kimes, D.S., Salas, W.A. and Routhier, M. (2000). Secondary forest age and tropical forest biomass estimation using Thematic Mapper imagery. Bioscience 50: 419–431.

Niklas, K.J. (1994). Plant Allometry. The Scaling of Form and Process. The University of Chicago Press, Chicago.

Niklas, K.J., (2004). Plant allometry: is there a grand unifying theorem? Biology Review 79: 871–889.

Niklas, K.J., Enquist, B.J. (2002). On the vegetative biomass partitioning of seed plant leaves, stems, and roots. American Naturalist 159: 482–497.

Pacala S.W., Hurtt G.C. and Baker D. (2001). Consistent land- and atmosphere-based US carbon sink estimates. Science 292: 2316–2320.

Parolin, P. (2002). Radial gradients in wood specific gravity in trees of Central Amazonian floodplains. IAWA Journal 23: 449-457.

Pilli, R., Anfodillo, T., Carrer, M. (2006). Towards a functional and simplified allometry for estimating forest biomass. Forest Ecology and Management 237: 583–593.

Popescu, S.C., Wynne, R.H., and Nelson, R.F. (2003). Measuring individual tree crown diameter with lidar and assessing its influence on estimating forest volume and biomass. Canadian Journal of Remote Sensing 29: 564–577.

Ochoa-Gaona, S. (2001). Traditional land-use systems and patterns of forest fragmentation in the highlands of Chiapas, Mexico. Environmental Management 27: 571-586.

Overman, J.P.M., Witte, H.J.L., Saldarriaga, J.G. (1994). Evaluation of regression models for above-ground biomass determination in Amazon rainforest. Journal of Tropical Ecology 10: 218–297.

Palacio-Prieto, J.L., Bocco, G., Velázquez, A., Mas, J.F., Victoria, A., Luna-González, L., Gómez-Rodríguez, G., López-García, J., Palma, M., Trejo-Vázquez, I., Peralta, A., Prado-Molina, J., Rodríguez-Aguilar, A., Mayorga-Saucedo, R. and González, F. (2000). La condición actual de los recursos forestales en México: resultados del Inventario Forestal Nacional 2000. Investigaciones Geográficas (43): 183-203.

Phillips, D. L., Brown, S.L., Schroeder, P.E., and Birdsey, R.A. (2000). Toward error analysis of large-scale forest carbon budgets. Global Ecology and Biogeography 9: 305-313.

Richardson, J., Bjorheden, R., Hakkila, P., Lowe, A.T., Smith, C.T. (2002). Bioenergy from Sustainable Forestry: Guiding Principles and Practice- Kluwer Academic Publishers. Dordrecht, The Neederlands. 344 p.

Rignot, E.J., Zimmerman, R. and Van Zyl, J.J. (1995). Spaceborne applications of P band imaging radars for measuring forest biomass. IEEE Transactions on Geoscience and Remote Sensing 33: 1162–1169.

Rojo-Martínez, G.E., Jasso-Mata, J., Vargas-Hernández, J., Palma-López, D.J., Velásquez-Martínez, A. (2005). Biomasa aérea en plantaciones comerciales de hule (Hevea brasiliensis Mull Arg.) en el estado de Oaxaca, México. Agrociencia 39: 449–456.

Saatchi, S.S., Houghton, R.A., Dos Santos-Alvala, R.C., Soarez, J.V., and Yu, Y. (2007). Distribution of aboveground live biomass in the Amazon Basin. Global Change Biology 13: 816-837.

Schumacher, F.X. and Hall, F.S. (1933). Logarthmic expression of timber-tree volume. J. Agric. Res. 47:719-734.

Schroeder, P., Brown, S., Birdsey, R., and Cieszewski, C. (1997). Biomass estimation for temperate broadleaf forests of the United States using inventory data. Forest Science 43: 424- 434.

Silva-Arredondo, M. and Návar-Cháidez J.J. (2009). Estimacion de factores de expansion de carbono en comunidades templadas del norte de Mexico. Revista Chapingo: Serie Ciencias Forestales y del Ambiente 15(2): 155-169.

Silva-Arredondo, M. and Návar J.J. (2010). Estimacion de factores de expansion de biomasa en comunidades templadas del norte de Mexico. Ciencia Forestal en Mexico. In Press.

Shinozaki, K.K., Yoda, K., Hozumi, K., Kira, T. (1964). A quantitative análisis of plant form - the pipe flor model tehory. 1. Basic Analysis. Japan. Journal of Ecology 14: 97-105.

Sprugel, D.G. (1983). Correcting for bias in log-transformed allometric equations. Ecology 64: 209–210.

Ter Mikaelian, M.T., Korzukhin, M.D. (1997). Biomass equations for sixty-five North American tree species. Forest Ecology and Management 97: 1–24.

Trejo, I. and Hernández, J. (1996). Identificación de la selva baja caducifolia en el estado de Morelos, México, mediante imágenes de satélite. Investigaciones Geográficas 5: 11-18.

Trejo, I. and Dirzo, R. (2000). Deforestation of seasonally dry tropical forest: a national and local analysis in Mexico. Biological Conservation 94: 133-142.

West, G.B., Brown, J.H., Enquist, B.J. (1999). A general model for the structure and allometry of plant vascular system. Nature 400: 664–667.

West, G.B., Brown, J.H., Enquist, B.J. (1999a). The fourth dimension of live: fractal geometry and allometric scaling of organisms. Science 284: 167- 169.

West, G.B., Brown, J.H., Enquist, B.J. (2001). A general model for ontogenetic growth. Nature 413: 628–631.

Woods, K.D., Feiveson, A.H., Botkin, D.B. (1991). Statistical error analysis for biomass density and leaf area index estimation. Canadian Journal of Forest Research 21: 974–989.

Zheng, D., Rademacher, J., Chen, J., Crow, T., Bresee, M., Le Moine, J. and Ryu, S. (2004). Estimating aboveground biomass using Landsat 7 ETM+ data across a managed landscape in northern Wisconsin, USA. Remote Sensing of Environment 93: 402–411.

Zeide, B. (1998). Fractal analysis of foliage distribution in loblolly pine crowns. Canadian Journal of Forest Research 28: 106–114.

Zianis, D. and Mencuccini, M. (2004). On simplifying allometric analyses of forest biomass. Forest Ecology and Management 187: 311–332.

Zianis, D., Muukkonen, P., Mäkipää, R. and Mencuccini, M. (2005). Biomass and stem volume equations for tree species in Europe. Silva Fennica 4: 1-63.

# 3

# Remote Characterization of Biomass Measurements: Case Study of Mangrove Forests

Temilola E. Fatoyinbo[1] and Amanda H. Armstrong[2]
*[1]NASA Goddard Space Flight Center, Greenbelt, Md*
*[2]University of Virginia, Charlottesville, VA*
*United States*

## 1. Introduction

Accurately quantifying forest biomass is of crucial importance for climate change studies and forest conservation. By quantifying the amount of above and belowground biomass and consequently carbon stored in forest ecosystems, we are able to derive estimates of carbon sequestration, emission and storage and help close the carbon budget.

Mangrove forests, in addition to providing habitat and nursery grounds for over 1300 animal species, are also an important sink of biomass. Although they only constitute about 3% of the total forested area globally, their carbon storage capacity-in forested biomass and soil carbon- is greater than that of tropical forests (Alongi, 2002; Lucas et al, 2007). In addition, the amount of mangrove carbon- in the form of litter and leaves- exported into offshore areas is immense, resulting in over 10% of the ocean's dissolved organic carbon originating from mangroves (Dittmar et al, 2006).

The measurement of forest aboveground biomass is carried out on two major scales: on the plot scale, biomass is measured from field measurements, allometric equation derivation and measurements of forest plots. On the larger scale, the field data are used to calibrate remotely sensed data to obtain stand-wide or even regional estimates of biomass. Currently, biomass can be calculated using average stand biomass values and optical data, such as aerial photography or satellite images (Landsat, MODIS, IKONOS, etc.). More recent studies have concentrated on deriving forest biomass values using radar (JERS, SIR-C, SRTM, Airsar) and/or Lidar (ICEsat/GLAS, LVIS) active remote sensing to retrieve more accurate and detailed measurements of forest biomass. The implementation of a generation of new active sensors, such as UAVSAR and ALOS/PALSAR has prompted the development of new techniques of biomass estimation that use the combination of multiple sensors and datasets, to quantify past, current and future biomass stocks.

Focusing on mangrove forest biomass estimation, this book chapter has 3 main objectives:

1. To describe in detail the field methodologies used to derive accurate estimates of biomass in mangrove forests.
2. To explain how mangrove forest biomass can be measured using several remote sensing techniques and datasets.
3. To describe the measurement challenges and errors that arise in estimates of forest biomass.

## 2. Methodology

### 2.1 Measurement of biomass in the field

a. Deriving allometric equations of mangrove trees

To calculate the biomass of an entire forest stand, the biomass (or weight) of individual trees in the must be calculated and summed. There are three main methods used to calculate stand biomass:

The harvest method is a technique where all of the trees in are felled, cut into sections and components (such as trunk, bark, leaves, branches), dried and subsequently weighed. This method is very labor intensive when dealing with trees that weigh several tons (Brown, 1997; Komiyama et al., 2005, 2008) and cannot be reproduced on a large scale because all of the trees within a set area have to be felled.

The 'mean tree method' consists in the weighing of one (or several) trees considered to be average, and extrapolating the biomass to that of the entire stand. This method can only be used in plantations or other stands with trees of a homogeneous size.

The most common method of stand biomass retrieval is using allometric equations. The allometric equations are derived from selective sampling of trees that are representative of the size-classes found in a forest. These equations then estimate the whole or partial weight of the trees relative to the tree metrics, such as diameter at breast height (DBH) and tree height. These equations have to be both site and species-specific, as even within-species biomass allocation can vary greatly depending on the location.

Allometry implies that the size and growth rate of one part of a living organism is proportional to the size and growth rate of another. In the case of trees, allometric equations correlate tree diameter with height, leaf biomass, root biomass, branch biomass, etc. Allometric equations to estimate biomass and growth of mangroves have been developed for several decades. These equations are available and applicable for all of the structural forms of mangroves including dwarf trees (Ross et al., 2001) single–stemmed, and multi-stemmed tree forms (Komiyama et al, 2008, Clough et al., 1997; Dahdouh-Guebas and Koedam, 2006). In their paper on mangrove allometry, Komiyama et al. (2008) describe the current state of knowledge on mangrove biomass and productivity equations based on 72 published studies in great detail. In their 1993 paper, Saenger and Snedaker also reviewed 43 aboveground biomass equations of mangroves worldwide, to derive a single, global height-biomass and height-productivity equation. Studies by Soares and Schaeffer-Novelli, Ong et al. (2004) and Comley and McGuiness (2005) describe the available species and site-specific equations extensively. As opposed to the site and species specific equations, Chave et al. (2005) and Komiyama et al. (2005) have proposed the use of common allometric equations that are not dependent on either site or species. These equations are dependend on wood density, the static model of plant form and the Shinozaki pipe model (Shinozaki et al., 1964; Oohata and Shinozaki, 1979). These common equations are of the form:

Komiyama et al., 2005:

$AGB = 0.251\rho DBH^{2.46}$ $r2 = 0.98$, with $n = 104$, $Dmax = 49$ cm, Relative error between 3.99 % and 30.1 % (1)

Chave et al., 2005:

$AGB = \rho \times \exp[-1.39 + 1.980\ln(DBH) + 0.207(\ln(DBH))^2 - 0.02081(\ln(DBH))^3]$; standard error of 19.5 % (2)

or

$$AGB = \exp(-2.977 + \ln(\rho DBH^2H)) = 0.0509 \times \rho DBH^2H;$$ standard error of 12.5 % with $n = 84$, $D_{max} = 50$ cm (3)

Where *AGB* is Aboveground biomass, ρ is wood density, (available at http://www.worldagroforestry.org/sea/Products/AFDbases/WD/index.htm), *DBH* is diameter at breast height and *H* is height of the tree. Figure 1 shows the allometric equations developed by Chave et al. (2005) and Komiyama et al (2008) for *Avicennia marina* mangrove trees.

When comparing the common equations to site and species specific equations, Komiyama et al. (2008) found that the average error was within 10%, thereby showing that wood density may be a more important factor in the determination of biomass than site or species.

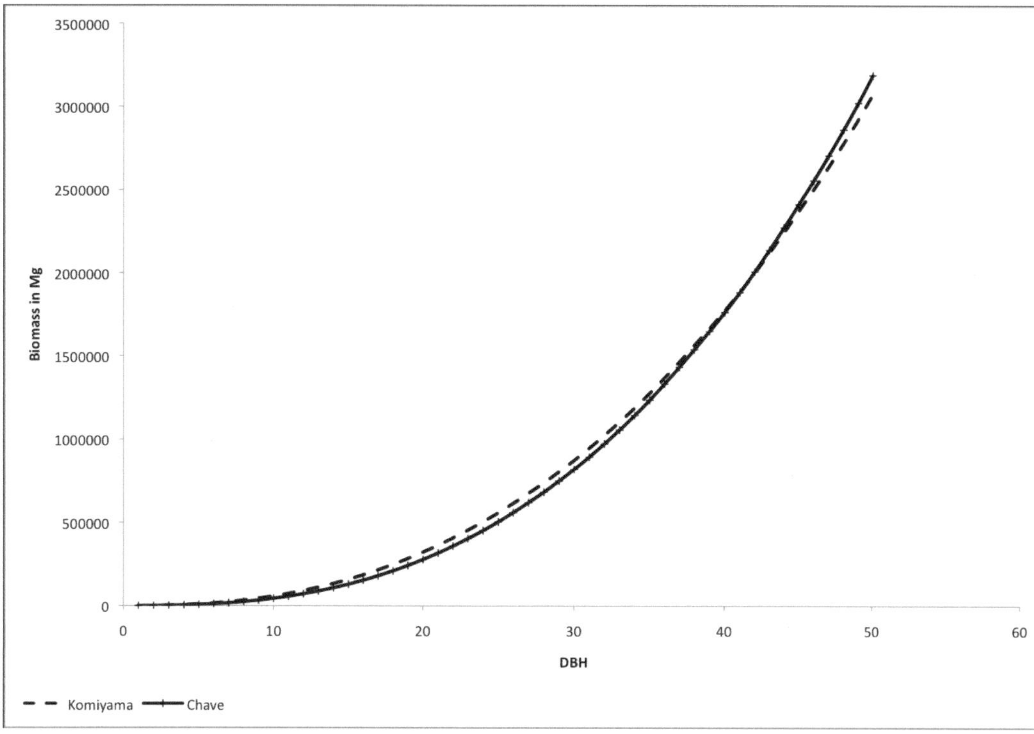


Fig. 1. Allometric equations developed by Chave et al (2005) and Komiyama (2008) for *Avicennia marina* trees. This plot shows the strong correlation between the two equations that are based on wood density.

b. Field Plots

The plot measurement method is the most common in situ approach to deriving stand level biomass. The philosophy behind this approach is that a representative sample of forest can intensively and non-destructively measured and then scaled up to derive forest-wide values. To begin with, an appropriate number of plots must be determined based upon the total size of the stand. Additionally, the plots must be located within the entire range of topography to capture as much local variation within the ecosystem as possible. One way to achieve a representative sampling of forest is to divide the forest into a grid and then establish and measure the same number of plots per grid square. The size of the grid would be determined by the total number of hectares that the study forest occupies.

Once the method for establishing study plots is decided upon, the size and shape of the study plot can be determined. Plots can be small in diameter and circular in shape, or large and rectangular depending on the amount of field records needed. Within each plot, GPS location should be recorded along with qualitative boundary descriptions to indicate location, canopy gap effects, proximity to water and other geographical variation. These qualitative descriptions can be useful in re-locating plots for additional study, and they can provide insight into explaining any drastic variation in biomass from stand-out plots. It is not uncommon to remove the 5% outlying plots with respect to biomass before scaling up plot data to arrive a forest-wide value.

After location data has been recorded, measurement of the trees in the plot can begin. From plot center, basal area and percent canopy cover is recorded. For all trees greater than a certain threshold (often 5cm) in diameter at breast height (DBH) the following is recorded: species identification, DBH; distance and direction from plot center, height of the tree. Trees with buttresses, aerial roots or similar features that preclude the measurement at breast height are measured just above the obstacle. Split trunk trees are treated as two trees if the split is below breast height and one tree if the split is above breast height. All regeneration 5cm- at breast height and smaller present in the plot is counted and the species are listed.

c. Wood density measurements

One of the largest challenges posed to the scientific community in understanding rainforest dynamics worldwide is the ability to accurately measure and analyze tree growth in an evergreen hardwood ecosystem. Few datasets exist on life histories of mangrove trees, as the ability to utilize dendrochronology techniques common to temperate forests is diminished by the absence of temperature driven seasonality in the tropics. In response to the need to understand how tree species in tropical forests worldwide grow, forest ecologists have developed alternative tools such as wood density analysis and cambial pinning techniques to measure mangrove tree species growth rates. Wood density has been determined to be an important physical characteristic of wood and it is related to other wood properties, including: resistance, porosity, and the number, size, and chemical composition of the cells (Noguiera et al 2005). In the tropics, wood density has been shown to relate to a tree's resistance to physical impacts caused by wind or strong tides to relative growth rate and mortality (Muller-Landau, 2004).

Density is measured and reported in a number of ways. Most commonly, wood density is measured from the wet and dry weights of small wood samples taken with an increment borer. The density is typically reported as the unit-less ratio. For tree biomass estimate derivation from forest volume data the appropriate density measurement is called 'basic density' or 'basic specific gravity', and is calculated as oven-dry weight divided by wet volume (Fearnside, 1997). The wet volume is achieved through soaking to saturation in the laboratory after sampling. This is because trees in a forest vary in moisture content depending on water availability, seasonal variation, competition and other physiological stress factors.

In order to sample for wood density, 12 mm increment borers are used to extract a small cylindrical sample that does not harm the sampled tree. Two samples are taken perpendicular to each other at or around 140 cm above the ground. Care must be taken when using an increment borer to core straight into the tree toward the pith, though only a 4-5cm sample is needed for density measurement. If the ratio of sapwood to heartwood is known for a given tree species, a further degree of accuracy can be achieved by coring to the pith and measuring sapwood and heartwood density separately. For most tropical tree species however, this ratio is unknown and therefore only one wood sample from each core (two per tree) need be obtained. Following extraction of wood samples, diameter and height of sampled trees will also need to be measured to calculate biomass. When in situ wood extraction has been completed, samples are soaked overnight in distilled water in a laboratory. Wet weights are measured to the nearest one thousandth milligram. Wood samples are then dried in an 80°C oven for 24 hours and dry weights are measured on the same scale.

### 2.2 Measurement of biomass from optical remote sensing

a. Using extent as a proxy for biomass

Optical or passive remote sensing uses visible and near-infrared reflectance from the earth to form images. This type of remote sensing data forms the basis for much of current global scale vegetation mapping due to the large number of sensors such as Landsat, MODIS, ASTER, IKONOS, etc., the greater ease of image interpretation and increasing numbers of freely available data archives. Google Earth ™ software for example, is based on a combination of optical remote sensing observations from MODIS, Landsat, Quickbird and in some instances aerial photography. Optical measurements have been widely used in studies that link AGB measurements from the field to satellite observations. The main challenge with optical data is the presence of persistent cloud cover, particularly in tropical regions, which make the use of optical data difficult.

The simplest approach to derive biomass from this type of data is to derive landcover or forest type using the optical data, then assign a value to each landcover type (in the case of mangroves these types could be determined by zonation, canopy shape, average density per pixel). To calculate biomass, the total area of each landcover type is then multiplied by the value. While this is the simplest method to estimate AGB, it does not take into account variations of structure and the error is great when looking at very large or very heterogeneous forests (Goetz et al, 2009).

b. Using NDVI as a proxy for biomass productivity

A variety of vegetation indices have been developed for retrieving vegetation structure from optical remote sensing. The most common way to estimate mangrove biomass is with the Normalized Difference Vegetation Index (NDVI) (Li et al 2007; Mather, 1999; Foody et al. 2001). The index is based on the characteristics that vegetation has noticeable absorption in the red and very strong reflectance in the near infrared (NIR). The formula used to calculate NDVI is (Mather, 1999):

$$NDVI = \frac{NIR - red}{NIR + red} \tag{4}$$

Different types of vegetation often show distinctive variability from one another due to such parameters as leaf shape, spacing of the plants, water content, and soil background. The use of NDVI has major drawbacks relating to biomass estimation - in addition to the problem arising from clouds, it has the problem of signal saturation at lower biomasses because of the shorter wavelengths that interact only with the canopy and do not take into account any effect of the trunk (Sader et al. 1989, Foody et al. 1996). Because the trunk is the main component of tree biomass, it is often underestimated. While the index has been shown to be effective in retrieving biophysical variables of temperate, low biomass vegetation (Foody et al. 2001), it has proven difficult to use in tropical rainforests and mangroves. According Li et al (2007), the NDVI can measure coastal area biomass with R= 0.626 and RMSE = 0.99 kg $m^{-2}$. However they found that using the optical data only significantly underestimated the biomass of some woody mangrove forests (e.g. *Sonneratia apetala*) because the NDVI reflects canopy properties rather than the trunk properties that are crucial for accurate biomass retrieval.

### 2.3 Measurement of biomass using active remote sensing

a. Polarimetric SAR

Synthetic Aperture Radar (SAR) uses microwaves emitted by an instrument and reflected by the earth to form an image. Polarized microwave signals can be horizontally (H) or vertically (V) transmitted and received, resulting in co (HH and VV) and cross (HV or VH) polarized data. In Polarimetric systems, the backscatter coefficient $\sigma^0$ (sigma nought in decibels dB) and phase can be derived for each polarization.

The backscatter coefficient of a forest canopy depends upon the interaction of microwaves with leaves, branches, trunks, and in the case of mangroves - aboveground roots. More specifically, the polarization, frequency and incidence angle of the microwaves and the size, density, orientation and dielectric constant of the vegetation components affect this backscatter coefficient $\sigma^0$ (Lucas et al. 2007). Longer wavelenths (L- and P- band) are able to penetrate the canopy and are scattered by larger components, such as the trunk and the ground and thereby increase the returned signal. Shorter wavelengths interact with leaves and twigs resulting in a larger amount of signal absorbed and less signal return.

Scattering refers to the interaction of microwaves with different surfaces and can range from direct scattering (mirror-like, when only one reflection occurs) to diffuse scattering (multiple reflections at different angles). In mangroves, the radar signature or backscattering coefficient can vary greatly depending on mangrove type and structure. In particular, increases in backscatter can be the result of:

- high surface roughness resulting from aboveground roots and large amount of dead wood material during low tides and dry season.
- double bounce effect resulting from scattering from tree trunks to the ground/water and back or scattering from the ground/water to tree trunks and back (MacDonald, 1980; Krohn et al, 1983; Imhoff, 1995; Simard et al 2002). This increases the backscattering coefficient beyond the saturation level (Mougin et al 1999; Proisy et al, 2002).

In forests, including mangroves, there is a positive relationship between measured backscattering coefficients $\sigma^0$ and the aboveground biomass. However this relationship only exists up to a threshold biomass value after which the backscattering coefficient saturates. The threshold is dependent on the polarization and wavelength of the radar signal. In mangroves, P-band frequency and HV polarization has been found to have the highest sensitivity to biomass, with a saturation level of 160 Mg ha -1, followed by L - HV (140 Mg/ha) and C-HV (70 Mg/ha) (Mougin, 1999; Proisy, 2002, Lucas, 2007).

The Japanese Aerospace Exploration Agency (JAXA) PALSAR (Phased Array L-Band Synthetic Aperture Radar) instrument on board the ALOS (ALOS (Advanced Land Observing Satellite)) platform is a fully polarimetric L-band SAR. The ALOS satellite was launched in 2006 and some preliminary studies have shown the use of L-band data to estimate forest biomass and structure (Lucas, 2007). The Kyoto and Carbon (K&C) Initiative initiated by the Earth Observation Research and Applications Center (EORC) in 2000 (Rosenqvist et al. 2003) has developed and validated products derived from the PALSAR sensor to address land cover (forest) mapping, forest change mapping and biomass and structure estimations (Lowry et al, 2010). These products are freely available at http://www.eorc.jaxa.jp/ALOS/en/kc_mosaic/kc_mosaic.htm.

Because mangrove aboveground biomass often exceeds the threshold of 140 Mg/ha that is measurable using backscatter alone, other techniques, where tree height is measured instead of biomass directly, are often more appropriate. To derive tree height, two different types of active sensors- Radar and Lidar- can be used alone, or in combination, to increase the area of coverage.

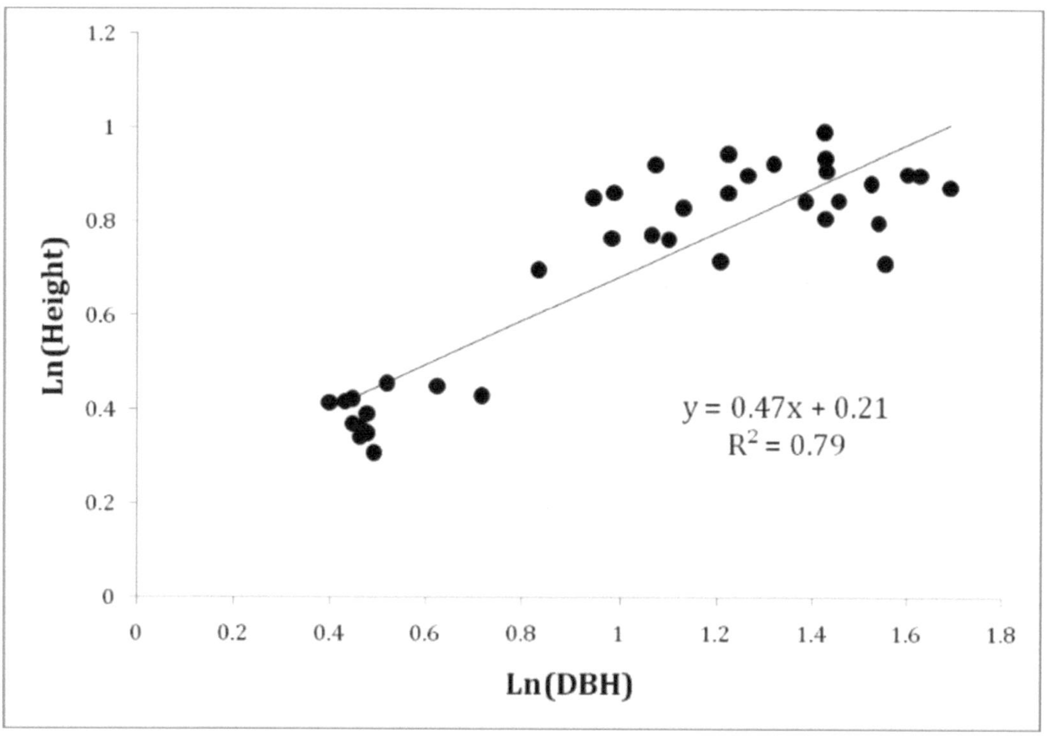


Fig. 2. Allometric equation relating the DBH to tree height in Avicennia marina trees. The equation relating DBH to height is ln(height) = 0.47*DBH +0.21, with an $R^2$ of 0.79.

b. Interferometric SAR

Forest structure (in terms of height and density) is a direct measurement that can be used to derive biomass, especially in high biomass systems, such as mangroves. To measure tree height using radar data, a technique known as interferometric Synthetic Aperture Radar (InSAR) is used (Graham, 1974). InSAR estimates the tree height by using interference patterns between two radar signals in order to derive terrain height. To derive biomass, the tree height is directly correlated to DBH (figure 2) and biomass through site-specific allometric equations or regional to global equations such as the one derived by Saenger and Snedaker (1993).

To quantify forest structure and make estimations of biomass in mangroves, the Digital Elevation Model (DEM) derived from the Shuttle Radar Topography Mission (SRTM) has proven most successful. In 2006 and 2008, Simard et al. used the SRTM DEM in combination with field validation data and Lidar to estimate mangrove forest 3-D structure and aboveground biomass. Fatoyinbo et al. (2008) used the combination of field data, Landsat and SRTM data to derive mangrove height, extent and aboveground biomass storage for Mozambique. The use of InSAR data, such as the SRTM DEM (or any other interferometric SAR dataset) to derive tree height is based on the principle that the radar signal measures the height at some depth in the tree canopy. In the case of C-band data, such as SRTM, the microwave signal penetrates the canopy and measures height at some depth within. X-band signals do not penetrate as deeply and measure height at a more shallow depth, whereas P- and L-band penetrate the canopy completely and measure deeper in the canopy than C band. Therefore, in order to measure the "true" height of the tree canopy, the DEM has to be calibrated by shifting the DEM height up to the "actual canopy height".

The height measurement that can be derived from InSAR data is the sum of the tree canopy height and the height of the ground. In forests where there is significant topography, the height of the ground has to be subtracted before calculating the height of the canopy. In mangroves however, the topography is negligible and the ground is considered flat, as these trees grow at sea level.

To calibrate the InSAR data, "real" canopy height measurements, from field measurements or Lidar data have to be used. Lidar (Light Detection and Ranging) measures vegetation height at very high accuracy (up to millimeters) and is considered the most accurate and consistent measurement of vegetation structure because of its systematic measurements and because field-based measurements are often limited in amount and spatial distribution. The ICESat/GLAS (Geoscience Laser Altimeter System) sensor is a spaceborne waveform Lidar system, which continuously records the amplitude of the lidar pulse returned through the different layers of the forest canopy. This provides a measurement of the vertical structure of the forest. The GLAS footprint has an ellipsoid form with a diameter of approximately 70m, and each footprint is separated by 172 m along track and 7.5 km between tracks (at the equator). Because the lidar only measures relatively few and small areas, it is generally used to calibrate other datasets. This data provides the best alternative for global canopy height calibration and is freely available from the National Snow and Ice Data Center (http://nsidc.org/data/icesat). An example of GLAS shots over the mangroves of the Niger Delta in Nigeria is presented in Figure 3 and in Figure 4 the mangrove height map of Nigeria and Cameroon derived from SRTM and GLAS is shown.

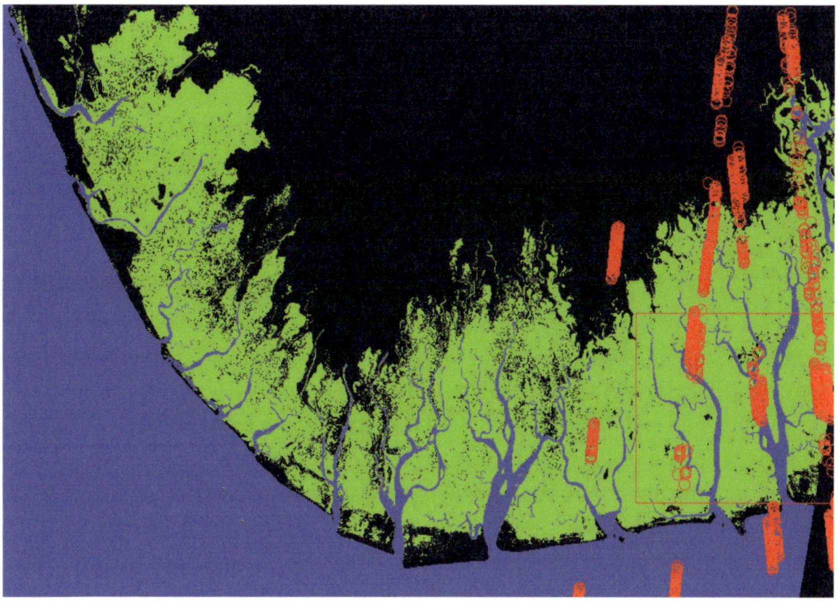

Fig. 3. Example of the available GLAS footprints available over the mangrove forests in the Niger Delta. The GLAS footprints are shown in red, mangrove forests in bright green, the ocean in blue and other landcover in black.

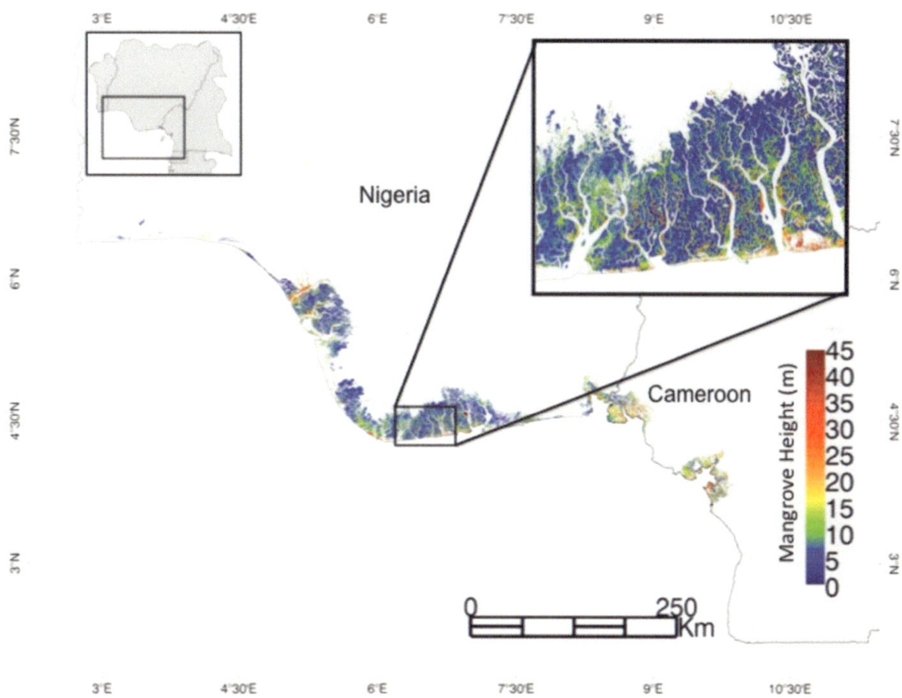


Fig. 4. Height map of mangrove forest in Nigeria and Cameroon derived from SRTM and GLAS.

## 2.4 Biomass measurement error

There are several types of error that can arise and need to be considered when working with remote sensing data to estimate biomass: For optical remote sensing, classification errors due to omissions and commissions, and clouds are the most prevalent. In Fatoyinbo et al (2008), we found that the Landsat-derived mangrove map did have a high accuracy of 93%, however, there was still some misclassification of mangrove areas as nonmangrove and vice versa, with 3.6% commissions and 10.6% omissions. For active remote sensing, a systematic error can be introduced from sensor error and there is also a certain amount of random error due to biases in measurements. Additional error is introduced by the utilization of different datasets: each dataset has a different resolution and was taken at a different time, which results in differences in measurement. Geolocation errors are introduced when using radar/lidar fusion of datasets in addition to error introduced due to the differing interaction of the radar and lidar signal with the canopy, soil and water in mangrove forests.

When using InSAR, Lidar and/or field data to estimate mangrove structure, there is also the possibility of geolocation error between the data. For example both SRTM and ICESat have mean geolocation errors of better than 20 m and 2.4 m +/- 7.4 m respectively (Carabajal, 2005). The error in geolocation can result in large differences in height measurements, which increases the total error. Furthermore, the natural variability of canopy height and structure

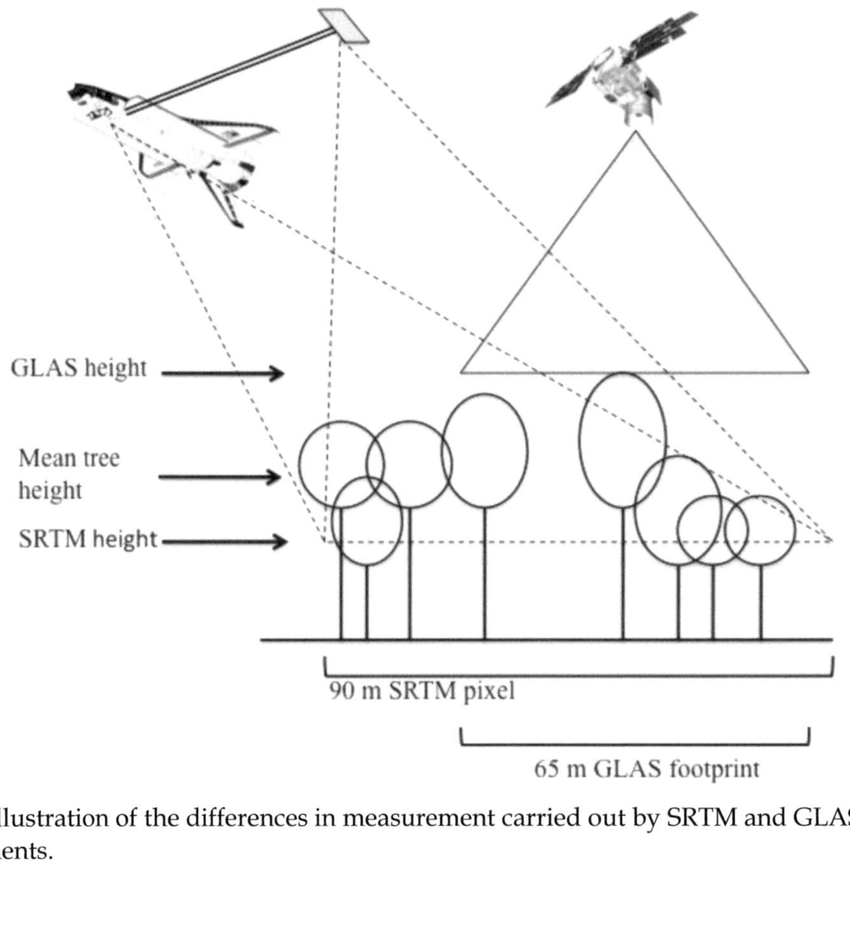


Fig. 5. Illustration of the differences in measurement carried out by SRTM and GLAS instruments.

within a forest increases possible measurement errors between the two datastets. This is compounded by the natural variability of canopy height within a forest. Therefore, if the trees measured by each method are not exactly the same, the differences between the height measurements and therefore the error of the measurement can be high. The differences in physical parameters measured by radar and lidar, in addition to differences in resolution also increase the height and biomass estimation error. In their 2006 study in the Florida Everglades, Simard et al. were able to calculate mean tree height within 2.0 m RMSE and Fatoyinbo et al. (2008) were able to estimate mean tree height within 1.6 m RMSE. Although this RMSE is very good when looking at forests at a whole, this methodology is not applicable to very short forests as the error can in this case be larger than the actual tree height. The combinations of sources of error are illustrated in figure 5.

Finally, when estimating the actual biomass of the mangroves from tree height, a large error can be introduced. The global height- biomass equation developed by Saenger and Snedaker (1993) for example, has a root mean square error of 65.4 Mg/ha due to the large variability of methodologies used to derive biomass in the dataset. Because the equation is applicable globally, it does not take into account local variations in species composition, height and biomass, thereby introducing potential error into the biomass estimate.

## 3. Conclusions

Measurement of aboveground biomass in forest ecosystems, including mangroves, is important for Carbon storage and cycling studies, mitigation of climate change and management of natural resources. In recent years, accurately quantifying biomass and carbon storage has become increasingly important for financial mechanisms of carbon emission mitigation such as Reduced Emissions from Deforestation and Degradation (REDD). In particular the UNFCCC and IPCC are pushing for increased large scale forest monitoring and development of carbon assessment methodologies. In this chapter, we highlight how field and remote sensing data can be used to estimate mangrove forest biomass. In particular, we concentrate on field measurement techniques and the application of active remote sensing using radar and lidar to better estimate mangrove height and biomass. While estimates of mangrove biomass have been achieved, even on a large scale, using different field and remote sensing techniques, challenges still remain. In particular, the potentially large error introduced by the combination of multiple datasets is a challenge when trying to estimate biomass with a low error. We therefore look forward to future satellite missions where radar and lidar data will be recorded simultaneously, such as the proposed NASA DesDynI mission.

## 4. References

Alongi, D. M. (2002), Present and future of the world's mangrove forests, Environ. Conserv., 29(3), 331–349.

Brown, Estimating biomass and biomass change of tropical forests: A primer, FAO For. Pap. 134, Food and Agric. Org., Rome.

Carabajal, C. C., and D. J. Harding (2005), ICESat validation of SRTM C-Band digital elevation models, Geophys. Res. Lett., 32, L22S01, doi:10.1029/2005GL023957.

Chave, J., Andalo, C., Brown, S., Cairns, M.A., Chambers, J.Q., Eamus, D., Fo □lster, H., Fromard, F., Higuchi, N., Kira, T., Lescure, J.P., Nelson, B.W., Ogawa, H., Puig, H., Rie´ra, B., Yamakura, T., 2005. Tree allometry and improved estimation of carbon stocks and balance in tropical forests. Oecologia 145, 87–99.

Clough, B.F., Dixon, P., Dalhaus, O., 1997. Allometric relationships for estimating biomass in multi-stemmed mangrove trees. Aust. J. Bot. 45,

Comley, B.W.T., McGuinness, K.A., 2005. Above- and below-ground biomass, and allometry of four common northern Australian mangroves. Aust. J. Bot. 53, 431–436.

Dahdouh Guebas, F., Koedam, N., 2006. Empirical estimate of the reliability of the use of the Point-Centered Quarter Method (PCQM): solutions to ambiguous field situations and description of the PCQM + protocol. Forest Ecol. Manage. 228, 1–18. Saenger, P., Snedaker, S.C., 1993. Pantropical trends in mangrove above-ground biomass and annual litterfall. Oecologia 96, 293–299.

Dittmar, T., Hertkorn, N., Kattner, G. and Lara, R.J. (2006). Mangroves, a Major Source of Dissolved Organic Carbon to the Oceans. *Global Biogeochemical Cycles* 20 (1), GB1012, doi:10.1029/1005GB002570

Fatoyinbo, T., et al., Landscape scale height, biomass and carbon estimation of mangrove forests with Shuttle Radar Topography Mission elevation data, *Journal for Geophysical Research-Biogeosciences*, 113, G02S06, 2008.

Foody, G.M., Cutler, M.E., Mcmorrow, J., Pelz, D., Tangki, H., Boyd, D.S. And Douglas, I., 2001, Mapping the biomass of Bornean tropical rain forest from remotely sensed data. Global Ecology and Biogeography, 10, pp. 379- 386.

Goetz, S.J., Baccini, A., Laporte, N.T., Johns, T., Walker,W., Kellndorfer, J., Houghton, R. A., and Sun, M. (2009) Mapping and monitoring carbon stocks with satellite observations: a comparison of methods. *Carbon Balance and Management.*

Graham, L.C., Synthetic interferometer radar for topographic mapping, Proceedings of the IEEE, 62, 763, 1974.M.

Imhoff ML. 1995a. A theoretical analysis of the effect of forest structure on synthetic aperture radar backscatter and the remote sensing of biomass. IEEE Transactions on Geoscience and Remote Sensing 33: 341–352.

Komiyama, A., Poungparn, S., Kato, S., 2005. Common allometric equations for estimating the tree weight of mangroves. J. Trop. Ecol. 21, 471–477

Komiyama, A., Ong, J.E., Poungparn, S., 2008. Allometry, biomass, and productivity of mangrove forests: A review. Aquat. Bot. 89, 128–137.

Krohn MD, Milton NM, Segal DB. 1983. Seasat synthetic aperture radar (SAR) response to lowland vegetation types in eastern Maryland and Virginia. Journal of Geophysical Research 88: 1937–1952.

LI, X, A. GAR-ON YEH, S. WANG, K. LIU, X. LIU, J. QIAN and X. CHEN, 2007. Regression and analytical models for estimating mangrove wetland biomass in South China using Radarsat images. International Journal of Remote Sensing Vol. 28, No. 24, 20 December 2007, 5567–5582

Lowry, J. L. Hess, et al. (2009). Mapping and Monitoring Wetlands Around the World Using ALOS PALSAR: The ALOS Kyoto and Carbon Initiative Wetlands Products: 105-120.

Lucas, R.M., et al., 2007. The potential of L-band SAR for quantifying mangrove characteristics and change: case studies from the tropics. *Aquatic Conservation: Marine and Freshwater Ecosystems*, 17, 245.

MacDonald HC. 1980. Techniques and applications of imaging radar. In Remote Sensing in Geology, Siegel BS, Gillespie AR (eds). John Wiley: New York; Chapter 10.

Mather, P.M., 1999, Computer Processing of Remotely-sensed Images, 2nd edn (Chichester: Wiley).

Mougin E, Proisy C, Marty G, Fromard F, Puig H, Betoulle JL, Rudant JP. 1999. Multifrequency and multipolarisation radar backscattering from mangrove forests. IEEE Transactions on Geoscience and Remote Sensing 37: 94–102.

Muller-Landau, H (2004). Interspecific and Inter-site Variation in Wood Specific Gravity of Tropical Trees. *Biotropica*, Vol. 36, No. 1 (Mar., 2004), pp. 20-32.

Nogueira, E. M., B. W. Nelson, et al. (2005). "Wood density in dense forest in central Amazonia, Brazil." Forest Ecology and Management 208(1-3): 261-286.

Ong, J.E., Gong, W.K., Wong, C.H., 2004. Allometry and partitioning of the mangrove, Rhizophora apiculata. Forest Ecol. Manage. 188, 395–408.

Oohata, S., Shinozaki, K., 1979. A statistical model of plant form—further analysis of the pipe model theory. Jpn. J. Ecol. 29, 323–335.

Ross, M.S., Ruiz, P.L., Telesnicki, G.J., Meeder, J.F., 2001. Estimating above- ground biomass and production in mangrove communities of Biscayne National Park, Florida (USA). Wetlands Ecol. Manage. 9, 27–37.

Shinozaki, K., Yoda, K., Hozumi, K., Kira, T., 1964. A quantitative analysis of plant form—the pipe model theory. II. Further evidence of the theory and its application in forest ecology. Jpn. J. Ecol. 14, 133–139.Li et al 2007

Simard M, De Grandi G, Saatchi S, Mayaux P. 2002. Mapping tropical coastal vegetation using JERS-1 and ERS-1 radar data with a decision tree classifier. International Journal of Remote Sensing 23: 1461–1474.

Simard, M., Zhang, K. Q., Rivera-Monroy, V. H., Ross, M. S., Ruiz, P. L., Castaneda-Moya, E., Twilley, R. R. and Rodriguez, E. (2006). Mapping Height and Biomass of Mangrove Forests in Everglades National Park with SRTM Elevation Data. Photogrammetric Engineering and Remote Sensing 72(3): 299-311.

Simard, M., Rivera-Monroy, V.H., Mancera-Pineda,J.E., Castañeda-Moya, E and R.R. Twilley (2008), A systematic method for 3D mapping of mangrove forests based on Shuttle Radar Topography Mission elevation data, ICEsat/GLAS waveforms and field data: Application to Ciénaga Grande de Santa Marta, Colombia. *Remote Sensing of Environment, Earth Observations for Terrestrial Biodiversity and Ecosystems Special Issue*, V. 112 (5), 2131-2144

Soares, M.L.G., Schaeffer-Novelli, Y., 2005. Above-ground biomass of man- grove species. I. Analysis of models. East. Coast Shelf Sci. 65, 1–18.,

Proisy C, Mougin E, Fromard F, Karam MA. 2000. Interpretation of polarimetric radar signatures of mangrove forests. Remote Sensing of Environment 71: 56–66.

Proisy C, Mougin E, Fromand F, Trichon V, Karam MA. 2002. On the influence of canopy structure on the polarimetric radar response from mangrove forest. International Journal of Remote Sensing 23: 4197–4210.Mougin, 1999;

Rosenqvist A, Shimada M, Igarashi T, Watanabe M, Tadono T, Yamamoto H. 2003. Support to multi-national environmental conventions and terrestrial carbon cycle science by ALOS and ADEOS-II}the Kyoto & Carbon Initiative. Proceedings of the International Geoscience and Remote Sensing Symposium (IGARSS '03), Toulouse, France, July 21-25, 2003. Institute of Electrical and Electronics Engineers (IEEE), Geoscience and Remote Sensing Society (GRSS), USA.

4

# On-line Biomass Estimation in a Batch Bio-technological Process: Bacillus Thuringiensis δ - Endotoxins Production.

Adriana Amicarelli, Fernando di Sciascio, Olga Quintero and Oscar Ortiz
*Universidad Nacional de San Juan-Instituto de Automática (INAUT)*
*Argentina*

## 1. Introduction

Biomass concentration in a biotechnological process is one of the states that characterize a bioprocess. Moreover, it is generally the main direct or indirectly desired product. It is well known that the biomass concentration is not normally measured online because this measurement is not possible or this is economically unprofitable. Therefore, for control purposes it is necessary to replace the unavailable biomass concentration measurements with reliable and robust online estimations. To this aim, several states observers can be found in the literature. A review of commonly used techniques can be found in (Bastin & Dochain, 1990; Dochain, 2003) and references therein. Observers can be coarsely divided into two broad classes: first principles or phenomenological estimators and empirical estimators. The phenomenological estimators can be also subdivided into classical observers and asymptotic observers. Classical observers include extended Kalman filter (EKF), extended Luenberger observer, high gain observer, nonlinear observers, and full horizon observer. In this class of estimators, a detailed knowledge of the reaction kinetics and associated transport phenomena are required to represent the balance equations. Modeling the biological kinetics reactions is a difficult and time-consuming task, and therefore the model used by the estimators could differ significantly from reality. This is the main disadvantage of these phenomenological estimators, i.e., their efficiency strongly relies on the model quality. Asymptotic observers are based on the idea that uncertainty in bioprocess models lies in the process kinetics models. The design of these observers is based on a state transformation performed to provide a model which is independent of the kinetics. A potential drawback of the asymptotic observers is that the rate of convergence is completely determined by the operating conditions, i.e., the rate of convergence can be very slow or the observer may not converge. Empirical estimators are based on constructing appropriate nonlinear models of biotechnological processes exclusively from the process input–output data without considering the functional or phenomenological relations between the bioprocess variables.

However, the conventional empirical modeling approach is based on the knowledge of the structure (functional form) of the data-fitting model (in advance). This is a difficult task since it involves the heuristic selection of an appropriate nonlinear model structure from numerous alternatives.

For the machine learning community, the data-based modeling of the biomass concentration from a finite number of noisy samples (the training dataset) is a supervised learning problem. From this area, in recent years, the artificial neural network methodology has become one of the most important techniques applied to biomass estimation, e.g. (Leal, 2001; Li, 2003; Amicarelli *et al.*, 2006) and references therein. Neal's work on Bayesian learning for neural networks (Neal, 1996) shows that many Bayesian regression models based on neural networks converge to a class of probability distributions known as Gaussian Processes according as the number of hidden neurons tends to infinity. Furthermore, Neal argued that in the Bayesian approach for real-world complex problems, neural network models should not be limited to nets containing only a small number of hidden units. Neal's observation motivates the idea of replacing parameterized neural networks and work directly with Gaussian Process models for the high-dimensional applications to which neural networks are typically applied (Neal, 1997).

This Chapter addresses the problem of the biomass estimation in a batch biotechnological process: the *Bacillus thuringiensis* (*Bt*) δ-endotoxins production process, and presents different alternatives that can be successfully used in this sense. The development of the Chapter includes the design of various biomass estimators, namely: a phenomenological biomass estimator, a standard EKF biomass estimator, a biomass estimator based on ANN, a decentralized Kalman Filter and a biomass concentration estimator based on Bayesian regression with Gaussian Process.

Finally, conclusions about the estimators are presented and the results show the techniques for the *Bacillus thuringiensis* δ-endotoxins production process on the basis of experimental data from a set of various fermentations.

## 2. *Bacillus thuringiensis* δ-endotoxins production process

### 2.1 Bioprocess description

In the last years, due to environmental reasons the interest in biological agents for their use in ecological insecticides (bioinsecticides) has notably increased. *Bacillus thuringiensis* is one of the microorganisms most frequently studied as toxin producer. *Bt* is an aerobic spore - former bacterium which, during the sporulation; also produces insecticidal crystal proteins known as δ-endotoxins. It has two stages on its life span: a first stage characterized by its vegetative growth, and a second stage named sporulation phase. When the vegetative growth finalizes, the beginning of the sporulation phase is induced when the mean exhaustion point has been reached. Normally the sporulation is accompanied by the δ-endotoxin synthesis. After the sporulation, the process is completed with the cellular wall rupture (cellular lysis), and the consequent liberation of spores and crystals to the culture medium (Starzak & Bajpai, 1991; Aronson, 1993, Liu & Tzeng, 2000).

This research has been conducted with the same process and fermentation conditions as the work of Atehortúa *et al.* (2007). The microorganisms used in this work were *Bacillus thuringiensis serovar. kurstaki* strain 172-0451 isolated in Colombia and stored in the culture collection of Biotechnology and Biological Control Unit (CIB), (Vallejo *et al.*, 1999). The medium (CIB-1) contained: $MnSO_4.H_2O$ (0.03 $g.L^{-1}$), $CaCl_2.2H20$ (0.041 $g.L^{-1}$), $KH_2PO_4$ (0.5 $g.L^{-1}$), $K_2HPO_4$ (0.5 $g.L^{-1}$), $(NH_4)_2SO_4$ (1 $g.L^{-1}$), yeast extract (8 $g.L^{-1}$), $MgSO_4.7H_2O$ (4 $g.L^{-1}$) and glucose (8 $g.L^{-1}$). Growth experiments of the fermentation process with *Bacillus thuringiensis* were performed in a reactor with a nominal volume of 20 liters (Fig.1). The fermentations were developed with an effective volume of 11 liters of cultivation medium, and they were inoculated to 10% (v/v) with the microorganism *Bt* culture. The inoculum added consisted of a vegetative phase culture: 5 mL spore suspension with 1 $\cdot 10^7$ UFC/mL (stored at -20 °C) was used to inoculate a 500 mL flask containing 100 mL of CIB-1, and incubated with shaking at 250 rpm at 30 °C during 13 h. Fifty milliliters of this culture were aseptically transferred to each one of two 2 L flasks containing 500 mL of CIB-1 and incubated as above for 5 h. The pH medium was adjusted to 7.0 with KOH before its heat sterilization. Culture conditions at harvest are typified by 90% free spores and δ-endotoxins crystals.

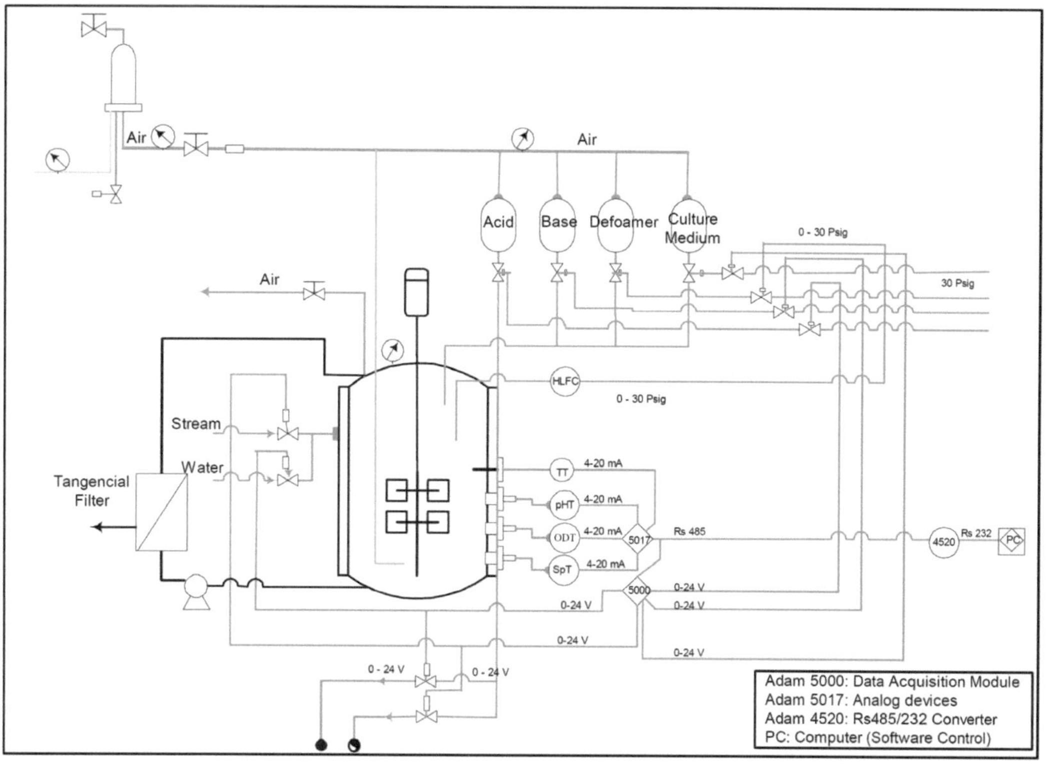


Fig. 1. Fermentation pilot plant scheme.

The temperature was maintained around 30 °C by using an ON/OFF control; whereas the pH was fixed between 6.5 and 8.5. The air flow was set up at 1320 [L.h$^{-1}$] and the agitation speed at 400 rpm. Manometric pressure in the reactor was set at 41,368 Pa using a pressure controller. Temperature, pH, dissolved oxygen, and glucose concentration were registered by a data acquisition system using an Advantech® PCL card. Dissolved oxygen was measured by a polarographic oxygen sensor InPro 6000 (Mettler Toledo, Switzerland), and glucose concentration was determined with a rapid off-line measurement method through a glucose analyzer (YSI 2700).

The reagents concentrations used for the pH control and foam formation were nitric acid (5N), potassium hydroxide (2N) and defoamer (33% v/v). Cell growth was determined as dry cell weight (Dry cell weight (DCW, g /L) = (final weight - initial weight)/(volume of microbial suspension filtered). The foam formation was avoided by manually aggregating a defoamer.

*Bacillus thuringiensis* δ-endotoxins production is an aerobic operation, i.e., the cells require oxygen as a substrate to achieve cell growth and product formation (Ghribi *et al.*, 2007).

### 2.2 Bioprocess model

The phenomenological estimator and the standard EKF presented in this Chapter are based on the phenomenological model presented in this Section, i.e. the model is necessary for its design. As pointed out in the introductory Section, the EKF is a classical nonlinear state estimator, and it's implemented for comparison purposes with the phenomenological biomass concentration estimator.

A first principle based model for *Bt* δ-endotoxins production process consists of a set of differential and algebraic equations (DAE system) in the continuous-time case, and a set of difference and algebraic equations in the discrete-time case. A simple phenomenological model was proposed by Rivera *et al.*, (1999), a modification to the Rivera model was given by Atehortúa *et al.*, (2006, 2007). Afterwards, Amicarelli *et al.* (2006, 2010) improved the model process adding the dissolved oxygen (DO) dynamics due to its importance in the biomass estimation problem and the posterior process control. The following state-space model is a discrete-time version of the continuous-time counterpart developed by Amicarelli *et al.* (2010).

$$\begin{bmatrix} X_v(k+1) \\ X_s(k+1) \\ S(k+1) \\ DO(k+1) \end{bmatrix} = \begin{bmatrix} \left(\left(\mu(k) - k_s(k) - k_e(k)\right)Ts + 1\right) X_v(k) \\ k_s(k) X_v(k)\, Ts + X_s(k) \\ -\left(\dfrac{\mu(k)}{Y_{x/s}} + m_s\right) X_v(k)\, Ts + S(k) \\ (K_1 - K_2 Ts)\, X(k) - K_1\, X(k+1) + DO(k) + K_3 Q_A\, Ts\left(DO^* - DO(k)\right) \end{bmatrix} \tag{1}$$

Where $X_v$ is the vegetative cell concentration, $X_s$ the sporulated cell concentration, $X = X_v + X_s$ is the total cell concentration ( $X(k+1) = \left(\mu(k) - k_e(k)\right) Ts X_v(k) + X(k)$ ), $S$ is the limiting substrate concentration and $DO$ is the dissolved oxygen concentration.

The following algebraic equations define the specific growth speed $\mu$ (model based on Monod equation for each limiting nutrient S and $DO$), the spore formation rate $k_S$, and the death cell specific rate $k_e$.

$$\mu(k) = \mu_{max}\left(\frac{S(k)}{\left(K_s + S(k)\right)}\frac{DO(k)}{\left(K_o + DO(k)\right)}\right) \tag{2}$$

$$k_s(k) = k_{smax}\left(\frac{1}{1+e^{Gs(S(k) - Ps)}}\right) - k_{smax}\left(\frac{1}{1+e^{Gs(S_{initial} - Ps)}}\right) \tag{3}$$

$$k_e(k) = k_{emax}\left(\frac{1}{1+e^{Ge(Tsk - Pe)}}\right) - k_{emax}\left(\frac{1}{1+e^{Ge(t_{initial} - Pe)}}\right) \tag{4}$$

The complete notation and model parameter's values are presented in Tables 1 and 2.

| Symbol | Description |
|---|---|
| $S$ | Limiting substrate concentration $\left[g.\,L^{-1}\right]$ |
| $Ts$ | Sampling time $[h]$ |
| $X_s$ | Sporulated cells concentration $\left[g.\,L^{-1}\right]$ |
| $X_v$ | Vegetative cells concentration $\left[g.\,L^{-1}\right]$ |
| $\mu$ | Specific growth rate $\left[h^{-1}\right]$ |
| $\mu_{max}$ | Maximum specific growth rate $\left[h^{-1}\right]$ |
| $m_s$ | Maintenance constant $\left[g\,substrate.\left[g\,cells.h^{-1}\right]^{-1}\right]$ |
| $k_s$ | Kinetic constant representing the spore formation $\left[h^{-1}\right]$ |
| $k_e$ | Death cell specific rate $\left[h^{-1}\right]$ |
| $Y_{X/S}$ | Growth yield $\left[g\,cells.g\,substrate^{-1}\right]$ |
| $K_s$ | Substrate saturation constant $\left[g.\,L^{-1}\right]$ |
| $K_O$ | Oxygen saturation constant $\left[g.\,L^{-1}\right]$ |
| $K_1$ | Oxygen consumption constant by growth (dimensionless) |
| $K_2$ | Oxygen consumption constant for maintenance $\left[h^{-1}\right]$ |
| $K_3$ | Ventilation constant $\left[L^{-1}\right]$ |
| $DO^*$ | O2 saturation concentration (DO concentration in equilibrium with the oxygen partial pressure of the gaseous phase) $\left[g.\,L^{-1}\right]$ |
| $Q_A$ | Air flow that enters the bioreactor $\left[L.\,h^{-1}\right]$ |

Table 1. Phenomenological model variables.

Four batch cultures with different initial glucose concentration (8, 21, 32 and 40 g.L$^{-1}$) were carried out to generate experimental data for model validation and parameters tuning. In this context, four parameter sets guarantee a representative covering of an intermittent fed batch culture (IFBC) with total cell retention (TCR) in the operation space according to the work of Atehortúa *et al.* (2007), see Table 2.
Maximum glucose concentration in the medium ($S_{max}$) was used as the switching criteria among the estimated batch parameter sets.

| | $S_{max}$ <10 g.L$^{-1}$ | 10g.L$^{-1}$< $S_{max}$ <20 g.L$^{-1}$ | 20g.L$^{-1}$< $S_{max}$ <32g.L$^{-1}$ | $S_{max}$ >32 g.L$^{-1}$ |
|---|---|---|---|---|
| $\mu_{max}$ [h$^{-1}$] | 0.8 | 0.7 | 0.65 | 0.58 |
| Yx/s [g. g$^{-1}$] | 0.7 | 0.58 | 0.37 | 0.5 |
| $K_s$ [g. L$^{-1}$] | 0.5 | 2 | 3 | 4 |
| $K_o$ [g. L$^{-1}$] | $1\times10^{-4}$ | $1\times10^{-4}$ | $1\times10^{-4}$ | $1\times10^{-4}$ |
| *ms* $\left[g.\left[g.h^{-1}\right]^{-1}\right]$ | $5\times10^{-3}$ | $5\times10^{-3}$ | $5\times10^{-3}$ | $5\times10^{-3}$ |
| ksmax [h$^{-1}$] | 0.5 | 0.5 | 0.5 | 0.5 |
| *Gs* $\left[g.L^{-1}\right]^{-1}$ | 1 | 1 | 1 | 1 |
| *Ps* [g.L$^{-1}$] | 1 | 1 | 1 | 1 |
| kemax [h$^{-1}$] | 0.1 | 0.1 | 0.1 | 0.1 |
| *Ge* [h$^{-1}$] | 5 | 5 | 5 | 5 |
| *Pe* [h] | 4 | 4.7 | 4.9 | 6 |
| $K_1$ dimensionless | $9.725\times10^{-4}$ | $4.502\times10^{-3}$ | $3.795\times10^{-3}$ | $1.597\times10^{-3}$ |
| $K_2$ [h$^{-1}$] | $1.589\times10^{-4}$ | $0.046\times10^{-3}$ | $0.729\times10^{-3}$ | $0.561\times10^{-3}$ |
| $K_3$ [L$^{-1}$] | $4.636\times10^{-4}$ | $0.337\times10^{-3}$ | $2.114\times10^{-3}$ | $1.045\times10^{-3}$ |
| *Ts* [h] | 0.1 | 0.1 | 0.1 | 0.1 |

Table 2. Model parameters for the intermittent fed batch culture with total cell retention of *Bacillus thuringiensis serovar. Kurstaki.*

### 3. Biomass concentration estimators design.

The duration of the batch fermentation is limited and depends on the initial conditions of the microorganism culture. All the fermentations used in this work were initialized with the same inoculate and different substrate concentration conditions (Atehortúa et al, 2007). When the medium is inoculated, the biomass concentration increases at expense of the nutrients, and the fermentation concludes when the glucose that limits its growth was consumed, or when 90% or more of cellular lysis is presented. After that, the latency period was removed (the bioprocess dead time is not considered), and the duration of each experiment is approximately 16 hours in this case.

The collected data from the fermentations is a set of concentrations measurements of dissolved oxygen (*DO*), primary substrate (*S*), and biomass (*X*) which have been sampled at different speed, 10 samples per hour for the concentrations of dissolved oxygen and glucose and 1 per hour for the biomass concentration, that was quantified by cell dry weight method. Practically, *DO* could be continuously measured whereas *S* can be measured up to 20 times per hour. From the bandwidth estimation of system signals by using Fourier frequency analysis, the sampling time $Ts = 1/10$ hours has been selected for dissolved oxygen and substrate measurements (di Sciascio & Amicarelli, 2008; Amicarelli, 2009).

In order to design biomass estimators for the *Bacillus thuringiensis* δ-endotoxins production process, it is proposed a two-stage method (di Sciascio & Amicarelli, 2008). In the first stage, the biomass concentrations data set is completed to have the same size as the dissolved oxygen concentration and primary substrate (glucose) concentration data sets. For this missing data problem (Little & Rubin, 2002), it was considered a Bayesian Gaussian Process Regression as an imputation strategy for filling the missing values. In the second stage, different biomass estimators are designed.

### 3.1 First stage design for all estimators- filling the biomass missing data

For the theory of Bayesian Regression Framework and Gaussian Process see Appendix C. Suppose that we have a noisy training data set *D* which consists of *m* pairs of *n*-dimensional input vectors $\{x_i\}$ (regression vector) joined in a n×m matrix **X**, and *m* scalar noisy observed outputs $\{y_i\}$ collected in a vector ***y***.

$$D = \left\{ \left( x_i, y_i \right) \middle| i = 1, L, m \right\} = \left\{ X, y \right\} \tag{5}$$

In order to construct a probabilistic statistical model for *D*, the following data-generating process is assumed:

$$y_i = f\left( x_i \right) + \varepsilon_i \tag{6}$$

where the latent real-valued function *f* is the deterministic or systematic component of the model, and the additive random term $\varepsilon$ is the observation error. The aim of regression is to identify the systematic component *f* from the empirical observations *D*.

In this section, the biomass concentration data vector is completed with virtual filtered measurements to have the same size as dissolved oxygen and substrate data vectors. This is a missing data problem, and the Gaussian Process Regression will be used as imputation method for filling the missing values (note that this task in a deterministic framework which can be viewed as a curve-fitting or interpolation problem).

For all experimental fermentations, the data-generating model for biomass concentration is:

$$X(tk) = \hat{X}(tk) + \varepsilon(tk) \tag{7}$$

The training data set $D$ consists of 18 pairs of time inputs $t = \{tk\} = \{1,...,18\}$ (in hours), and noisy biomass measurements outputs $X = \{X_k\} = \{X(t_1),...,X(t_{18})\}$. The latent functions $\hat{X} = \{\hat{X}_k\} = \{\hat{X}(t_1),...,\hat{X}(t_{18})\}$ are the estimated biomass concentrations.

The expression "Gaussian Process Regression Model" refers to the use of a Gaussian Process as a prior on **f**. This means that every finite-dimensional marginal joint distributions of function values **f** associated to any input subset of **X** is multivariate Gaussian.

$$p\left(f|X,\theta_P\right) = N\left(m(X), K\left(X,\theta_P\right)\right) \tag{8}$$

A Gaussian Process is fully specified by a mean function $m(X) = [m(x_1), L, m(x_m)]^T$ and a positive-definite covariance matrix $K(X, \theta_P)$, and it can be viewed as a generalization of the multivariate Gaussian distribution to infinite dimensional objects. Choosing a particular form of covariance function, the hyperparameters $\theta_P$ may be introduced to the Gaussian Process prior. Depending on the actual form of the covariance function $K(X, \theta_P)$ the hyperparameters $\theta_P$ can control various aspects of the Gaussian Process.
In this work, the elements of the parameterized covariance matrix, $C(X, \theta_P, \sigma^2)$, are denoted $C_{ij} = C(x_i, x_j)$, and they are functions of the training input data $\mathbf{X}$, because these data determine the correlation between the training data outputs $y$. A suitable parametric form of the covariance function is:

$$Cij = \theta_0 + \theta_1 \exp\left[-\frac{1}{2}\sum_{l=1}^{n}\frac{\left(x_i^{(l)} - x_j^{(l)}\right)^2}{r_l^2}\right] + \theta_2\,\delta(i,j) + \sum_{l=1}^{n}\alpha_l x_i^{(l)} x_j^{(l)} \tag{9}$$

where $x_i^{(l)}$ is the $l^{th}$ dimension of the input vector, $x_i$.
From the training data $D$, and by means of a conjugate gradient routine $\#\theta = 5$ hyperparameters, and the matrix **C** are determined recursively through:

$$\log\theta = [\log\theta_0, \log\theta_1, \log r_1, \ldots, \log r_n, \log\theta_2, \log\alpha_1, \ldots, \log\alpha_n]^T \tag{10}$$

and

$$\begin{cases} L = -\dfrac{1}{2}\log|C| - \dfrac{1}{2}y^T C^{-1} y - \dfrac{m}{2}\log 2\pi + \log p(\theta) + c \\ \dfrac{\partial L}{\partial\theta_i} = -\dfrac{1}{2}\operatorname{trace}\left(C^{-1}\dfrac{\partial C}{\partial\theta_i}\right) + \dfrac{1}{2}y^T C^{-1}\dfrac{\partial C}{\partial\theta_i}C^{-1}y + \dfrac{\partial\log p(\theta)}{\partial\theta_i} \end{cases} \tag{11}$$

Afterwards, at different times, $t_* = 0.1, 0.2, ..., 17.9, 18$ by (12)

$$\hat{f}_* = E\left(f_* \middle| D, x_*, K, \sigma^2\right) = k_*^T C^{-1} y$$
$$\sigma^2_{\hat{f}_*} = k_{**} - k_*^T C^{-1} k_* \qquad (12)$$

the latent functions $\hat{X}_* = \{\hat{X}_*\} = \{\hat{X}(t_*)\}$ and the variance $\sigma^2_{\hat{X}_*}$ are estimated. The expression "virtual filtered measurements" refers to the latent functions $\hat{X}_*$, because the additive normal noise $\varepsilon$ has been removed (filtered) from the "virtual measurement" $X_*$ in the data-generating model (7). Figure 2 gives an example of completion of biomass missing data for two fermentations (Fermentation 1, and Fermentation 2).

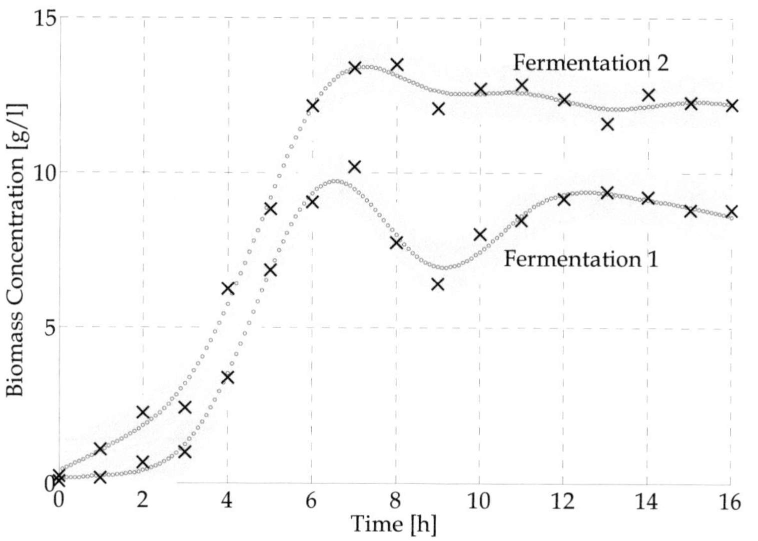


Fig. 2. Example of completion of biomass missing data for Fermentation 1 and Fermentation 2. The crosses being the biomass concentration measurements (training data), the small circles represent the biomass estimated (virtual filtered biomass measurements), and the grey region depicts the 95% confidence interval for the estimations (±2 standard deviations) (from di Sciascio & Amicarelli, 2008).

### 3.2 Phenomenological observer

In order to design a biomass phenomenological estimator, the dissolved oxygen balance from the nonlinear state-space model (1) presented before is employed in this Section

$$X(k) = \frac{1}{K_1}\left[\left(K_1 - K_2 Ts\right) X(k-1) - DO(k) + DO(k-1) + K_3 Q_A Ts\left[DO^* - DO(k-1)\right]\right] \qquad (13)$$

From (13) it can be inferred that online, the total biomass concentration can be estimated with experimental data of dissolved oxygen concentration (*DO*) and with biomass past values (X(k-1)) for the current estimation. The remaining constants and parameters are known for this estimator. As the biomass is normally measured using an offline method, in this case the dry weight method, the mentioned past values are not available for online estimation at the instant $k$. For this reason, it is not realistic to use the biomass measurements obtained by dry weight method and consequently, for online biomass estimation the values provided from the phenomenological model (1), $X(k-1) = Xv(k-1) + Xs(k-1)$ were used. Figure 3 shows the model structure for the phenomenological biomass estimator.

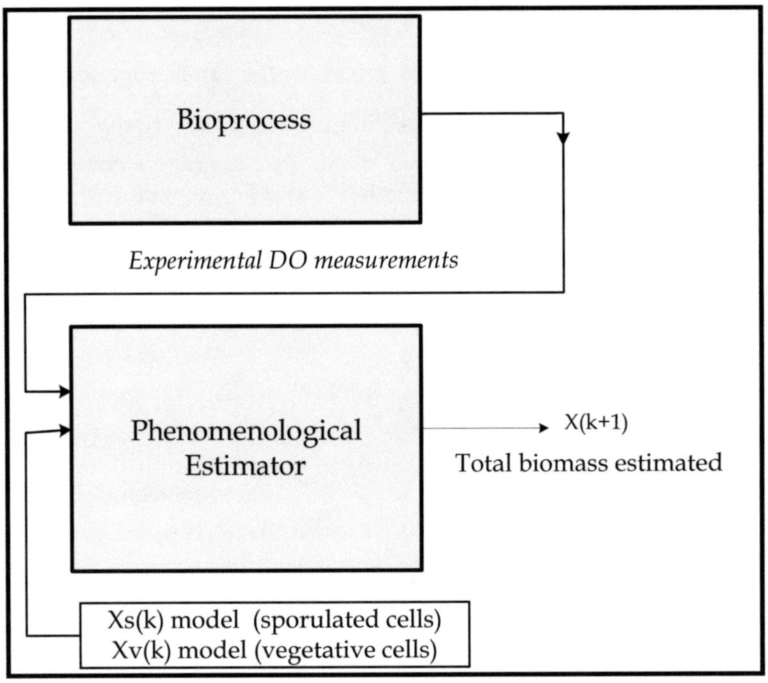


Fig. 3. Simulated output model structure for the phenomenological biomass estimator.

Figure 4 shows the phenomenological estimation results. This observer can approximate the biomass concentration better than the first model proposed by Atehortúa *et al.* (2007).
This is because, this estimator includes the dissolved oxygen consumption for growth and maintenance of the microorganism on its structure and through the experimental data of dissolved oxygen available online (Fig.3). In Fig 5. it can be seen the dissolved oxygen percentages time evolution for both fermentations.
Moreover, Fig. 4 shows satisfactory results and a correct behavior of the phenomenological estimator for two different fermentations. Estimated biomass follows closely the real biomass measurements. Similar results can be obtained for almost all fermentations. It can be noted that this performance is achieved by a phenomenological observer derived from the dissolved oxygen model available for this process. It is important to remark that the estimator involves in its structure the original model of vegetative and sporulated cells, whereas the consideration of the dissolved oxygen influence on the microorganism concentration improves the biomass estimation performance. It is important to remark that when the DO influence is not significant, the biomass estimation achieved with the model without the dissolved oxygen dynamics and the phenomenological estimator are comparable (show Fermentation 1 in Fig.4.). However, for those cases in which the DO approaches critical values (see Fermentation 2 in Fig. 4), the phenomenological observer gives better estimations (Fermentation 2).

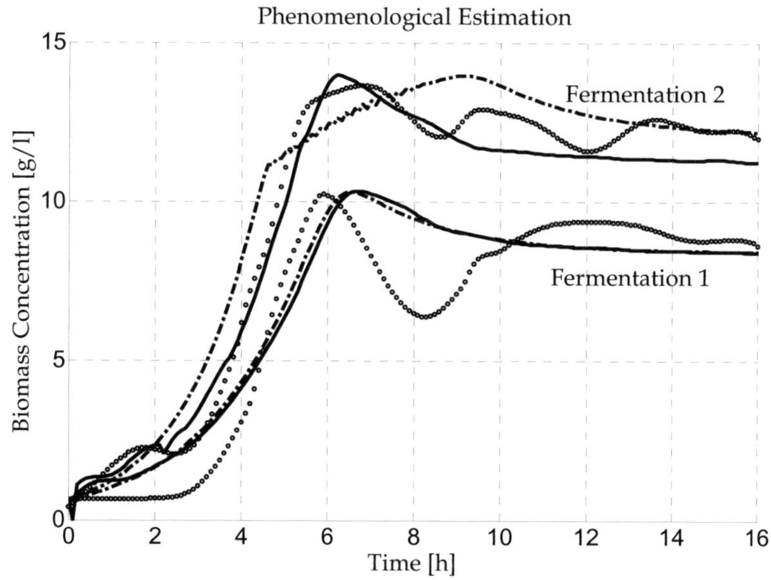


Fig. 4. Biomass estimator performance. The dash-dot line describes the behavior of biomass when considering the model (1); the solid line depicts the phenomenological estimator behavior based on DO dynamics; and the real biomass measurements are represented by small circles.

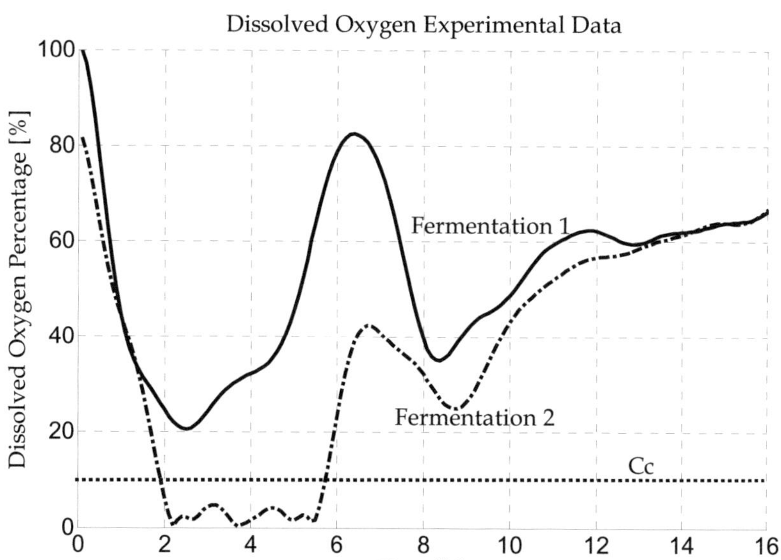


Fig. 5. Dissolved Oxygen experimental data. The solid line describes the Dissolved Oxygen behavior for the Fermentation 1; the dash-dot line depicts the Dissolved Oxygen behavior for the Fermentation 2 and the dotted line corresponds to the percentage of Dissolved Oxygen for the critical Dissolved Oxygen concentration for this process.

### 3.3 EKF standard estimator

Kalman filters are a widely useful tool used in biomass estimation due to its convergence and reliability properties. The estimation achieved from Kalman filters depends on the process model accuracy as well as on the available state measurements. The suitability of this estimation method can be concluded for the *Bt* fermentation process. Furthermore, this Section proposes a biomass concentration estimator for the mentioned biotechnological batch process through an Extended Kalman Filter (EKF) implementation.

The underlying theory of the EKF is largely known in the literature devoted to filtering, estimation, and control; see, for example, the classic books by Jazwinski (1970), Anderson & Moore (1979), or most recently, the book by Simon (2006). Therefore, in this work only brief explanations of the specific EKF implementation are given. In the EKF framework, the state transition and observation models are nonlinear differentiable states functions.

State transition model:

$$x(k+1) = f(x(k), u(k), k) + w(k) \tag{13}$$

Measurements model:

$$y(k) = h(x(k), k) + v(k) \tag{14}$$

Where $f(\times,\times)$ is the state transition function; $h(\times,\times)$ is the measurement function; $x(k)$ is the system state vector with initial condition $x(0) \sim N(x_0, Q_0)$ (as is usual in statistical literature the symbol (~) means "distributed according to"); $u(k)$ is the input or control vector; $y(k)$ is the observation vector; $w(k)$ is a discrete-time normal white noise process (process noise) with null mean and covariance matrix $\boldsymbol{Q}$, i.e., $w(k) \sim N(0, Q)$; and $v(k)$ is a discrete-time normal white noise process (measurements noise) with null mean and covariance matrix $\boldsymbol{R}$, i.e., $v(k) \sim N(0, R)$. The initial condition $x(0)$, and the sequences $w(k)$, and $v(k)$ are uncorrelated for all time shifts.

In our case the nominal State transition model (without the process noise $w(k)$) is obtained by introducing (2), (3) and (4) in (1).

$$x(k+1) = f(x(k), k) \tag{15}$$

The system state vector is $x(k) = [X_V(k) \quad X_S(k) \quad S(k) \quad DO(k)]^T$, the input vector is $u(k) = 0$ (the bioprocess has no external input), and the bioprocess outputs (observation vector) is $y(k) = [S(k) \quad DO(k)]^T$ (Fig.7). The experimental dissolved oxygen percentages and substrate concentration data employed are shown in Fig. 5 and 6.

The measurement model is linear in the states:

$$y(k) = Hx(k) \tag{16}$$

where $H = \begin{bmatrix} 0 & 0 & 1 & 0 \\ 0 & 0 & 0 & 1 \end{bmatrix}$

Taking into account the scales of the outputs, a balanced linear combination of $S(k)$ and $DO(k)$ can be considered as an alternative measurement model.

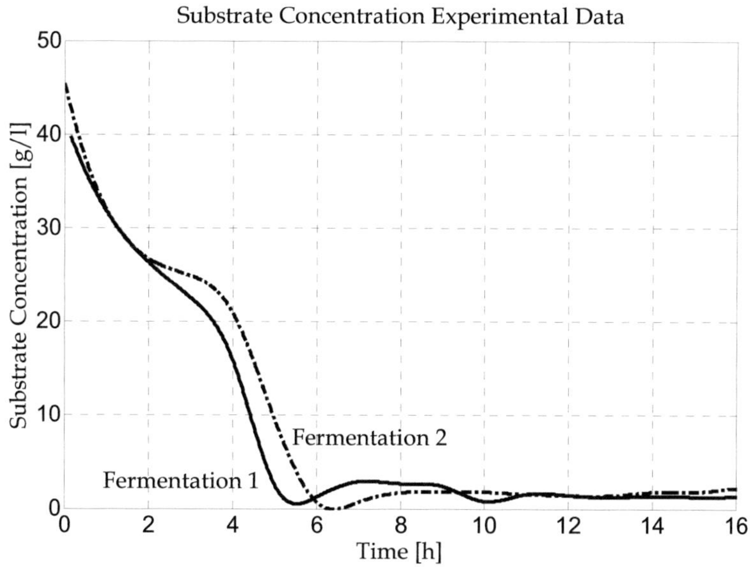


Fig. 6. Substrate Concentration experimental data. The solid line describes the Substrate Concentration for the Fermentation 1 and the dash-dot line depicts the Substrate Concentration for the Fermentation 2.

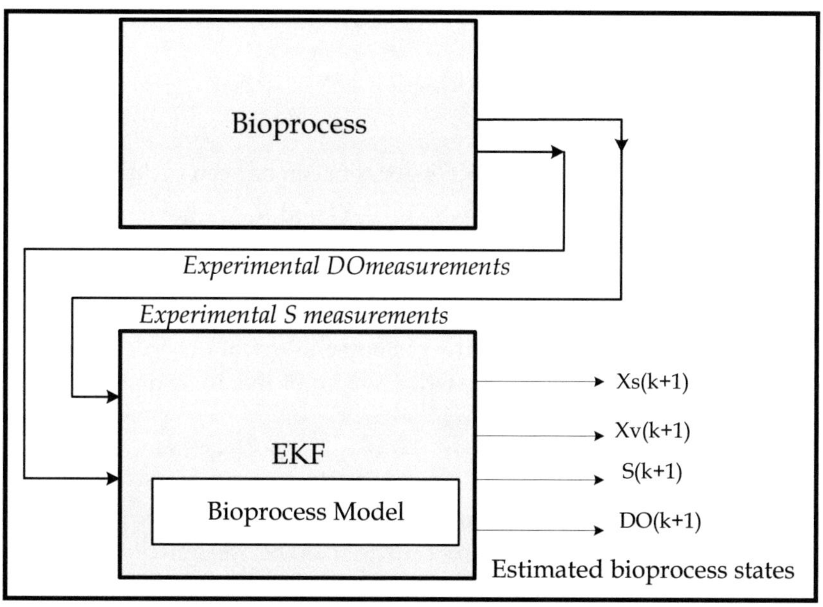


Fig. 7. Simulated output model structure for EKF biomass estimator.

$$y'(k) = H'x(k) = \alpha S(k) + \beta DO(k) \tag{17}$$

In this measurement model $H' = [0 \quad 0 \quad \alpha \quad \beta]$

where:

$$\alpha = DOmax/(S\ max + DOmax) \qquad \beta = S\ max/(Smax + DOmax)$$

The next step is to obtain the Jacobian matrices $\frac{\partial f(x(k),k)}{\partial x}$, and $\frac{\partial h(x(k),k)}{\partial x}$ evaluated at $\hat{x}(k-1|k-1)$.

$$A(k) = \left.\frac{\partial f(x(k),k)}{\partial x}\right|_{\hat{x}(k-1|k-1)}$$

$$A(k) = \begin{bmatrix} \left.\frac{\partial f_1(x(k),k)}{\partial x_1}\right|_{\hat{x}(k-1|k-1)} & \cdots & \left.\frac{\partial f_1(x(k),k)}{\partial x_4}\right|_{\hat{x}(k-1|k-1)} \\ \cdot & \cdot & \cdot \\ \left.\frac{\partial f_4(x(k),k)}{\partial x_1}\right|_{\hat{x}(k-1|k-1)} & \cdots & \left.\frac{\partial f_4(x(k),k)}{\partial x_4}\right|_{\hat{x}(k-1|k-1)} \end{bmatrix} \tag{18}$$

$$H(k) = \left.\frac{\partial h(x(k),k)}{\partial x}\right|_{\hat{x}(k|k-1)} = \left.\frac{\partial H x(k))}{\partial x}\right|_{\hat{x}(k|k-1)} = H \tag{19}$$

The entries of the matrix $A(k)$ and the EKF algorithm can be seen in Appendix A.

Finally, initializing the elements of the matrices $P$, $Q$ and $R$, we have all the components of the EKF algorithm (see Table 3 in Appendix A). In order to obtain the best possible fit of the EKF to the experimental data, the elements of the matrices $Q$ and $R$ are empirically adjusted by simulations. Figure 8 shows results for two different fermentations. It is performed a comparison between this estimator and the phenomenological observer based on dissolved oxygen dynamics (DO) previously presented. The aim of this investigation is to remark the relevance of the information used for both observers.

It can be concluded that the performance of the standard EKF estimator is adequate. This of course does not mean that the performance of the EKF cannot be meaningfully enhanced by using a better model of the bioprocess or by some of the numerous improvements to the basic EKF scheme. In particular, different EKFs can be designed using a long list of engineering tricks: different coordinate systems; different factorizations of the covariance matrix; combinations of all of the above, as well as other bells and whistles invented by engineers in the hope of improving higher order Taylor series corrections to the state vector EKF performance (Daum, 2005).

The phenomenological estimator presents an adequate behavior, but their efficiency strongly relies on the model quality for this dissolved oxygen dynamics. It should be noticed that both estimators highlight the importance of the DO dynamics for this process.

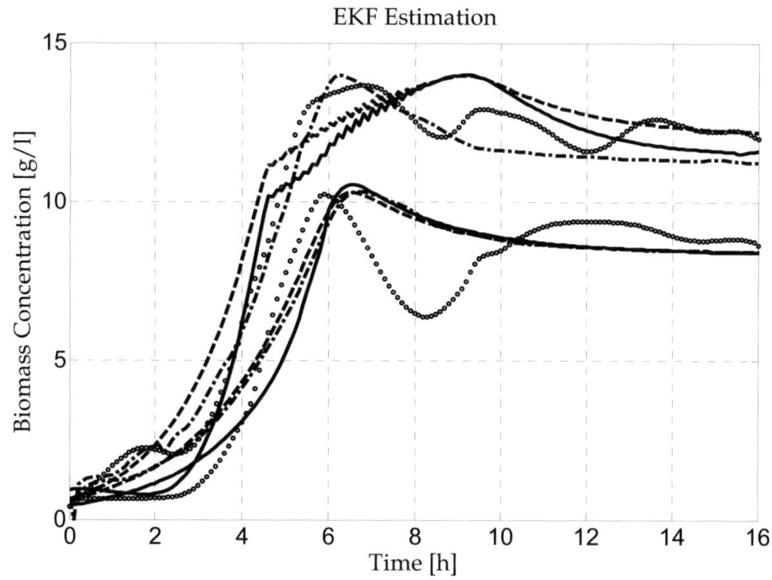


Fig. 8. Biomass estimator performance. The dashed line describes the biomass evolution obtained from the original model (1); the solid line depicts the EKF behavior, the dashed-dotted line depicts the phenomenological estimator behavior based on DO dynamics; and the real biomass measurements are represented by small circles.

### 3.4 ANN based estimator

Through artificial neural networks (ANN) the empirical knowledge (set of measurements) that characterizes a phenomenon of interest can be adequately codified. Due to the high degree of parallelism, the high generalization capability and the possibility to use an architecture of multiple inputs and outputs, the ANNs can provide a satisfactory solution to the problems of models identification, variables estimation, pattern recognition, functions approximation, among others. ANNs have the ability to abstract automatically essential characteristics of the experimental data, and to generalize from the previous experience; this allows the identification of the model process at lower cost.

Supervision and control techniques require optimizing the fermenter operation and the monitoring of all variables online is the best solution, since the methods offline delay the possibility of getting results and generally require more effort.

The ANN employed in this work is a multilayer perceptron with a hidden layer of 30 neurons and one output layer. For the training stage the Back Propagation algorithm (Haykin, 1999) was employed. The network was trained with data from a fermentation identified as "Fermentation 1" (See Fig. 9) and was generalized with other set of experimental data "Fermentation 2" (See Fig. 10). The activation functions of the hidden layer were hyperbolic tangent and a linear function for the output layer.

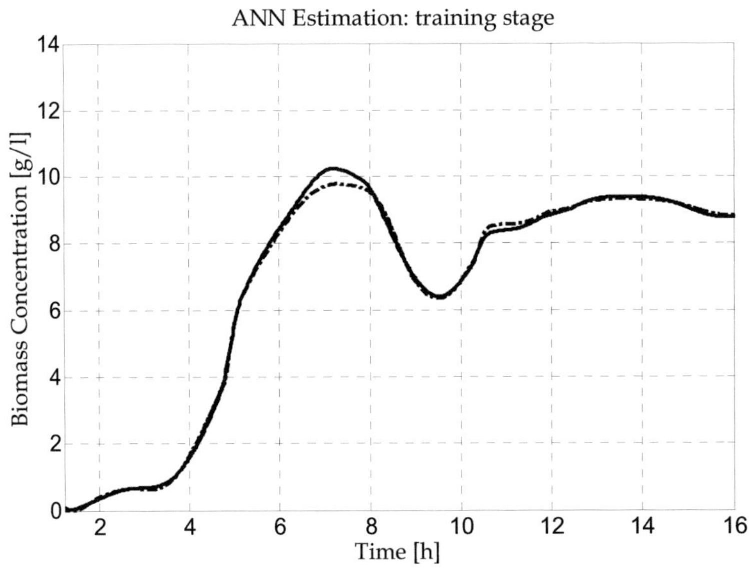


Fig. 9. Biomass estimator performance. The dashed line describes the biomass evolution obtained by the ANN in the training stage and the real biomass measurements are represented by the solid line. The perceptual training error e= 0.16 %.

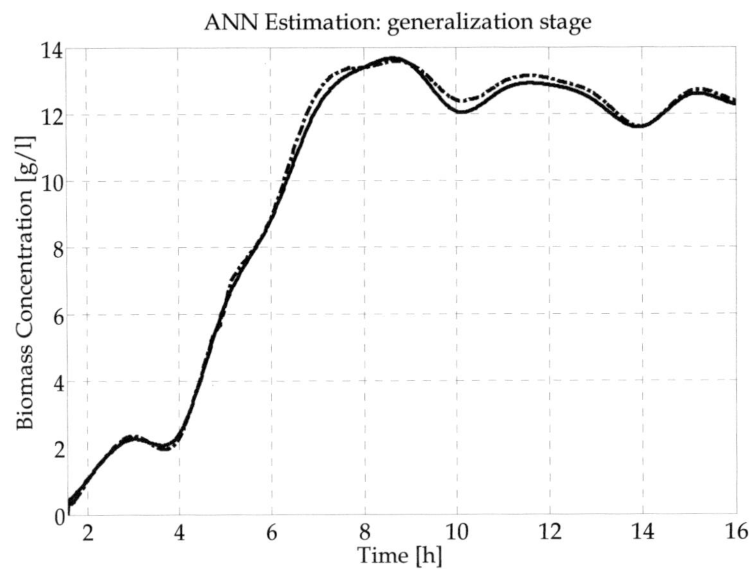


Fig. 10. Biomass estimator performance. The dashed line describes the biomass evolution obtained from the ANN in the generalization stage and the real biomass measurements are represented by the solid line. The perceptual generalization error e= 0.25 %.

### 3.5 Fusion through decentralized Kalman filter

The aim of this Section is to obtain an optimal biomass value for the process of *Bt*. To do this, two measurements (estimates) sequences are considered: the biomass estimation available from the phenomenological observer and the biomass estimation provided by the ANN-based observer. Assuming that the estimations are the optimum value for each sequence in time and the relationship between these values is given by:

$$X^{i} = X^{iOPT} + v^{i} \tag{20}$$

where $v^{i}$ is a random variable with zero mean and covariance $R^{i}$. In order to obtain an optimum value for biomass estimation, it was considered a decentralized Kalman filter (Brawn, 1997). In a basic approach of the decentralized Kalman Filter, each local filter operates autonomously. Each local filter has its own set of measurements, and there is no sharing of measurements. Note that this is inherently a cascaded operation mode, because the outputs of one or more of the local filters are acting as inputs to the master filter. The local filters (one for each sequence of measurements), the master filter and the different variables involved can be appreciated in Fig. 11.

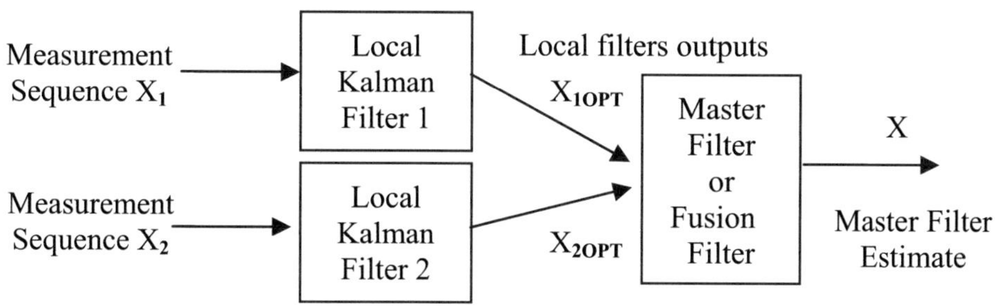


Fig. 11. Fusion scheme through a Decentralized Kalman Filter.

The mean and covariance for each sequence of measurements are calculated recursively according to:

$$\hat{X}^{i}(k+1) = \hat{X}^{i}(k) + \mu \left( X^{i}(k) - \hat{X}^{i}(k) \right) \tag{21}$$

$$R^{i} = R^{i} + \mu \left( \left( X^{i} - \hat{X}^{i} \right)^{2} - R^{i} \right) \tag{21}$$

where $\hat{X}^{i}$ is the average sequence value of $X^{i}$ and $0 < \mu < 1$ is a design constant. Then each sequence is individually filtered:

$$\left(P^{i}\right)^{-1} = \left(M^{i}\right)^{-1} + \left(R^{i}\right)^{-1} \tag{23}$$

$$X_{V}^{i.OPT} = P^{i} \left[ m^{i} \left(M^{i}\right)^{-1} + \left(R^{i}\right)^{-1} X_{V}^{i} \right] \tag{24}$$

Equation (23) provides the updated information matrix and Eq. (24) are the states estimated updates, $M^{i}$ and $m^{i}$ are the covariance error and the previous estimation values for the measurements sequences $X^i$ respectively. All values are merged to obtain the optimum value of the estimated biomass. In Fig. 12 the results achieved with this approach can be observed.

$$X = P\left[\frac{m}{M} + \sum_i \left(\frac{X_1^{i.OPT}}{P^i} - \frac{m^i}{M^i}\right)\right] \tag{25}$$

$$P^{-1} = M^{-1} + \sum_i \left[\left(P^i\right)^{-1} - \left(M^i\right)^{-1}\right] \tag{26}$$

$$\begin{aligned} M^i &= P^i \\ m^i &= X^{i.OPT} \\ m &= X \\ M &= P \end{aligned} \tag{27}$$

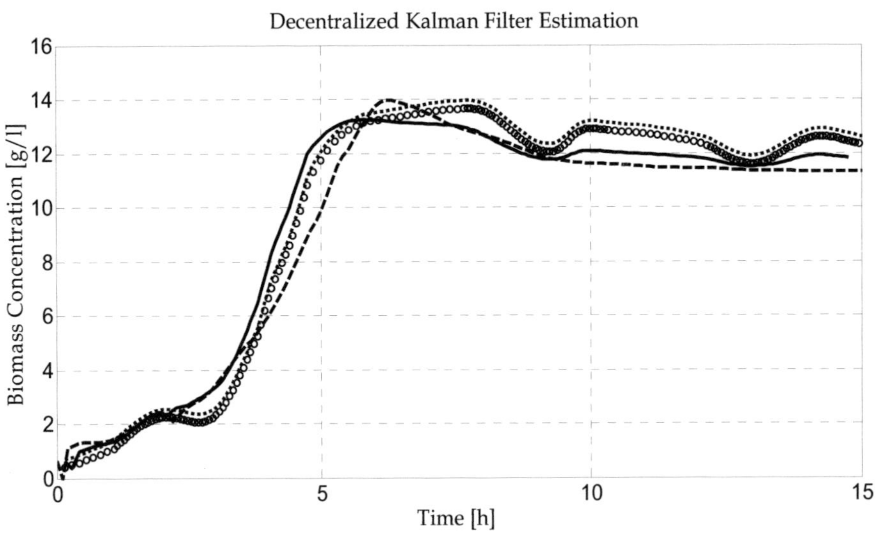


Fig. 12. Biomass estimator performance. The dashed line describes the biomass evolution obtained from the phenomenological estimator and the dot line describes a biomass estimation obtained from the ANN. The solid line describes the biomass evolution obtained through the decentralized Kalman Filter. The real biomass measurements are represented by small circles.

This architecture allows the complete autonomy of the local filters. The system achieves optimality in each individual local filter and global optimality in the primary filter.

### 3.6 Estimator based on Bayesian Regression through Gaussian Process

The first step in this design, is selecting the regressors, i.e., the components of the input vector $\boldsymbol{x}$. This is a laborious task, and has been done heuristically, chosen from numerous alternatives. The best empirical results have been achieved with:

$$x(kTS) = [DO(kTs), S(kTs), \hat{X}((k-1)Ts), \hat{X}((k-2)Ts)]^T$$

where $k = \{1, L, 180\}$ is the time index, $Ts = 1/10$ hours is the sampling time, $DO(.)$ is the dissolved oxygen concentration, $S(.)$ is the substrate concentration, and $\hat{X}(.)$ is the virtual filtered biomass measurement. In this case, the training data set $D$ consists of 180 pairs of input vectors $\{x(kTs)\} = \{x_k\} \in R^4$ collected in a matrix $X \in R^{4\times180}$, and scalars outputs $\{\hat{X}(kTs)\} = \{\hat{X}_k\}$ collected in a vector $\hat{X} \in R^{180}$ (note that in this section, the virtual filtered biomass measurements $\{\hat{X}_k\}$ are considered as true observed measurements). For the theory of Bayesian Regression Framework and Gaussian Process see Appendix C.

The data-generating process is $\hat{X}_k = \hat{\hat{X}}_k + \varepsilon_k$, being the latent function $\hat{\hat{X}}_k(.)$, and the additive normal noise $\varepsilon$. Once again, the $\#\theta = 11$ hyperparameters, and the new covariance matrix **C** eq. (9) are determined via a conjugate gradient routine from (11) and:

$$Cij = \theta_0 + \theta_1 \exp\left[-\frac{1}{2}\sum_{l=1}^{n}\frac{\left(x_i^{(l)} - x_j^{(l)}\right)^2}{r_l^2}\right] + \theta_2\,\delta(i,j) + \sum_{l=1}^{n} \alpha_l\, x_i^{(l)} x_j^{(l)} \qquad (28)$$

Furthermore, by (12) the biomass concentration $\hat{\hat{X}}_* = \hat{\hat{X}}(t_*)$ and the variance $\sigma^2_{\hat{\hat{X}}_*}$ are estimated for a set of different times $\{t_*\}$, $0 < t_* < 18\,\text{hours}$.

For the training stage, a one-step ahead predicted output schema is performed, i.e., the input measurements, $DO(k)\ S(k)$, and the previous output measurements $\hat{X}(k-1)$, $\hat{X}(k-2)$ are used as repressors in:

$$\hat{\hat{X}}(k) = \hat{\hat{X}}(DO(k), S(k), \hat{X}(k-1), \hat{X}(k-2))$$

For on-line estimation the implemented estimator is the simulated output schema, i.e., only input measurements DO(k), S(k) are used. The simulated output is obtained as above, by replacing the measured outputs by the simulated output from the previous steps, i.e., previous outputs from the model have to be fed back into the model computations on-line (Fig.13).

$$\hat{\hat{X}}(k) = \hat{\hat{X}}(DO(k), S(k), \hat{\hat{X}}(k-1), \hat{\hat{X}}(k-2))$$

The one-step ahead predicted output scheme is also known as Nonlinear Auto Regressive with Exogenous input model (NARX), or as Series-Parallel model. Furthermore, the simulated output schema is known as Nonlinear Output Error model (NOE), or as Parallel model (Narendra & Parthasarathy, 1990; Ljung, 2006).

The biomass concentration of fermentations Fermentation 1 and Fermentation 2 from the preceding section (see Fig.2) has been adopted as training, and validation data respectively. Figures 5 and 6 show the measurements of dissolved oxygen percentages (*DO*) and glucose concentration (*S*) respectively. Both signals have been filtered with a low-pass filter with a 1/36 Hz corner frequency.

Figure 14 shows the results for the proposed biomass estimator and the results from the previous section, i.e., the true biomass measurements, the virtual filtered biomass measurements, and the 95% confidence intervals.

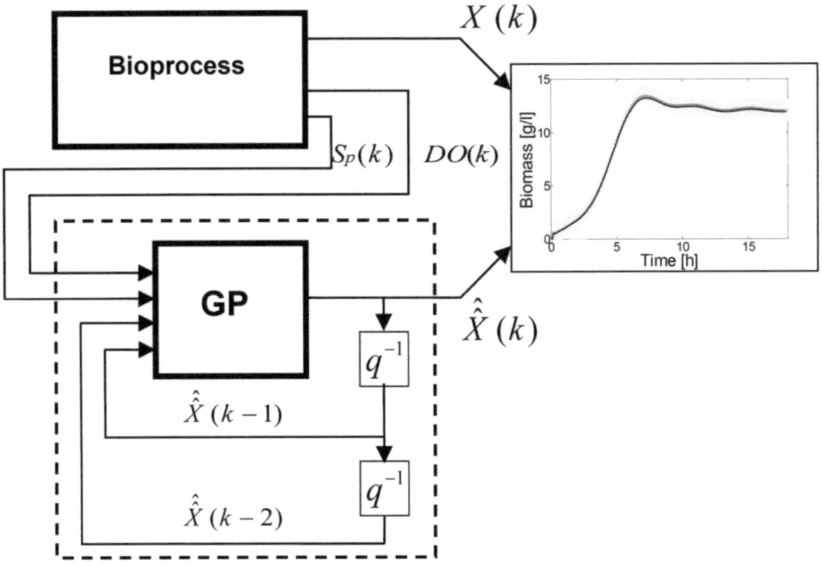


Fig. 13. Simulated output model structure of proposed biomass estimator (from di Sciascio & Amicarelli, 2008).

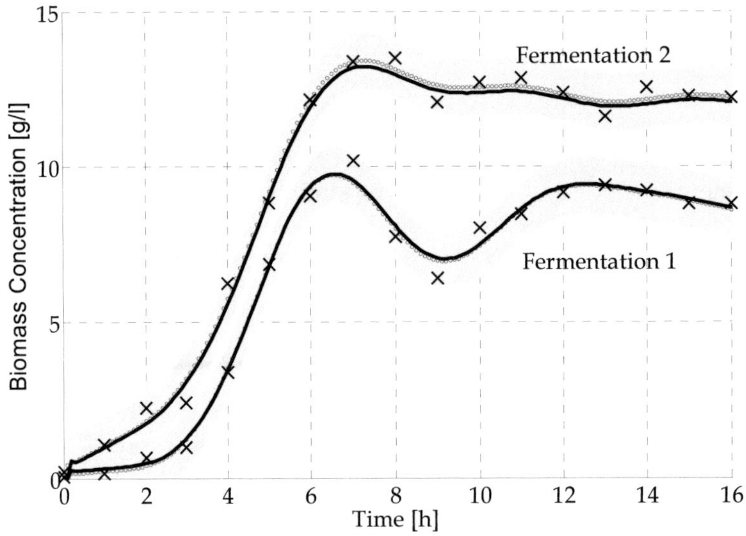


Fig. 14. Biomass estimator performance. The bold solid-line describes the behavior of the proposed biomass estimator ( $\hat{\hat{X}}(kTs)$ ), the crosses are the true biomass measurements, the virtual filtered biomass measurements ( $\hat{X}(kTs)$ ) are represented by small circles, and the grey region depicts the 95% confidence interval (from di Sciascio & Amicarelli, 2008).

From Fig. 14, the correct behavior of the proposed biomass estimator can be clearly seen. The estimated biomass follows closely the true and the virtual filtered biomass measurements, and similar results can be obtained for almost all (except for some atypical) fermentations. This performance is achieved settling the 11 hyperparameters of the covariance function. For example, in order to obtain a similar performance with a multilayer feedforward neural network with one hidden layer (Haykin, 1999; Bishop, 1995) it can be shown that more than 30 neurons are necessary (Amicarelli *et al.*, 2006; Amicarelli, 2009). This means that hundreds of parameters must be calculated during the training phase.
The most probable explanation for poor results in some atypical fermentations is the aggregation of antifoam throughout the fermentations; this operation can be viewed as an unmodelled perturbation. If the dosage isn't correct, i.e., too much defoamer is aggregated, then DO concentration decreases markedly, and due to the aerobic nature of the *Bacillus thuringiensis* production process, this affects the estimator performance adversely. This undesirable behavior of the estimator can be avoided or at least minimized by an adequate control of the antifoam dosage.
The time evolution of the biomass concentration is considered as an uncertain dynamic system perturbed by process noise, i.e., a stochastic process, and the evolution of biomass concentration for a particular fermentation is a realization of the stochastic process. Furthermore, *Bacillus thuringiensis* has two stages on its life span (vegetative growth and sporulation) with very different dynamics clearly distinguishable in Fig. 14. This means that the time evolution of biomass concentration is a non-stationary stochastic process, moreover the last term of the covariance function (9) capture the non-stationarity of the biomass evolution.

## 4. Discussion and conclusions

In this Chapter it has been addressed the problem of the biomass estimation in a batch biotechnological process: the *Bacillus thuringiensis* (*Bt*) δ-endotoxins production process. Different alternatives that can be successfully used in this sense were presented. It has been exposed the design of various biomass estimators, namely: a phenomenological biomass estimator, a standard EKF biomass estimator, a biomass estimator based on ANN, a decentralized Kalman Filter, and a biomass concentration estimator based on Bayesian regression with Gaussian Process.
Each estimation method has its own advantages and drawbacks according to their ability to take into account the model uncertainties and the measurement errors. For all the proposed estimators, at the first design stage, the biomass concentrations data set was completed to have the same size as in dissolved oxygen percentage and substrate (glucose) concentration data sets.
The phenomenological biomass estimator is based on information from the dissolved oxygen balance for the process. The biomass concentration does not depend only on the mentioned variable; therefore, the proposed observer does not take into account biomass variations produced due to variations of temperature, pH, antifoam aggregation and inherent conditions of inocula. This estimator has achieved satisfactory results and a correct behavior for a set of different fermentations. This performance was reached by a phenomenological observer derived from a dissolved oxygen model available for this process. This observer considered the DO influence on the microorganism concentration, which can improve the biomass estimation performance when the DO influence is significant.

The observation vector of the standard extended Kalman Filter observer was built with experimental dissolved oxygen and substrate data. This estimator also provides an adequate biomass estimation. However the phenomenological estimator has a better behavior. On the other hand, the phenomenological estimator efficiency strongly relies on the model quality for this process. It should be noted that both estimators highlight the importance of dissolved oxygen for this process and are based on the quality of the process model.
The ANN based estimator was a Recurrent Multilayer Perceptron. The proposed virtual sensor provided satisfactory results for the biomass estimation showing acceptable performance. The choice of the input variables is important, since it is a batch process which has "infinite memory". Next, the result provided by the phenomenological and the ANN observers was compared with a new estimation given by its fusion through a Decentralized Kalman Filter. This new estimation is useful when redundant and comparable information from different sensors exists.
In designing a biomass concentration estimator based on Bayesian regression with Gaussian Process; the time evolution of biomass is conceived as a dynamic system perturbed by a certain process noise. Although the bioprocess is not truly stochastic, this noise is used for modeling the uncertainties in the system dynamics, i.e., the stochasticity is only used for representing the model uncertainties. Biomass concentration estimation is obtained indirectly through observed noisy measurements. Noise in the measurements refers to a disturbance in the sense that the measurements are uncertain, i.e., even in the hypothetical case that the true biomass concentration is known, the measurements would not be deterministic functions of this true biomass, but would have a certain distribution of possible values. The major difficulty when the biomass estimation is implemented is related to the uncertainty of the models used to describe their dynamics.
The proposed biomass concentration estimator based on Bayesian regression with Gaussian Process was formulated like a filtering problem. In the design, a Gaussian Process Regression schema was used for biomass estimation. The regressors selection and the estimator stability are important. Regressors selection is a problem related to the choice of states in a state space representation of the system. Finding a set of "good" regressors for biomass estimation is a non trivial task, because this set is characteristic of each specific bioprocess. In this work, this task has been done heuristically by trial and error between numerous alternatives; however, more elaborated methods can be used, for example, optimization-based regressors selection, Analysis of Variance (ANOVA), and so on (Hastie *et al.*, 2001; Lind, 2006; Mannale, 2006). The first option is the set of regressors $\{DO(kTs), DO((k-1)Ts), \hat{X}((k-1)Ts)\}$ that can be derived from the phenomenological model of Section 2.2, (Amicarelli *et al.*, 2006, Amicarelli, 2009). However, the best empirical results have been achieved with the regressors $\{DO(kTs), S(kTs), \hat{X}((k-1)Ts), \hat{X}((k-2)Ts)\}$.
As a result of the fed back of preceding outputs, the proposed on-line biomass estimator is a dynamic system, therefore potential instability problems can appear, even if the original off-line predicted estimator is stable. It is very difficult to analyze analytically the on-line estimator stability properties and only simulations analyses have been carried out previously to implementation. In this specific application such potential instability problems have never appeared.
It is important to remark that, the performance achieved with the Bayesian estimator is achieved setting 11 hyperparameters of the covariance function and in order to obtain a similar performance with a multilayer feed-forward neural network it can be shown that at least 30 neurons are necessary.

## Appendix A. Extended Kalman Filter implementation

The entries of the matrix A(k) are:

$$A(k) = \left.\frac{\partial f(x(k),k)}{\partial x}\right|_{\hat{x}(k-1|k-1)}$$

$$A(k) = \begin{bmatrix} \left.\frac{\partial f_1(x(k),k)}{\partial x_1}\right|_{\hat{x}(k-1|k-1)} & \cdots & \left.\frac{\partial f_1(x(k),k)}{\partial x_4}\right|_{\hat{x}(k-1|k-1)} \\ . & . & . \\ \left.\frac{\partial f_4(x(k),k)}{\partial x_1}\right|_{\hat{x}(k-1|k-1)} & \cdots & \left.\frac{\partial f_4(x(k),k)}{\partial x_4}\right|_{\hat{x}(k-1|k-1)} \end{bmatrix}$$

$$a_{11} = 1$$

$$a_{12} = \frac{k_{smax}Ts}{1+e^{Gs(S(k)-Ps)}} - \frac{k_{smax}Ts}{1+e^{Gs(S_{initial}-Ps)}}$$

$$a_{13} = -\frac{k_{smax}TsX_v(k)Gs\,e^{Gs(S(k)-Ps)}}{\left(1+e^{Gs(S(k)-Ps)}\right)^2}$$

$$a_{14} = 0$$

$$a_{21} = 0$$

$$a_{22} = 1 + \frac{k_{emax}Ts}{1+e^{-Ge(t_{initial}-Pe)}} - \frac{k_{emax}Ts}{1+e^{-Ge(Tsk-Pe)}} + \frac{k_{smax}Ts}{1+e^{Gs(S_{initial}-Ps)}} - \frac{k_{smax}Ts}{1+e^{Gs(S(k)-Ps)}} + \frac{TsS(k)DO(k)\mu_{max}}{(K_d+DO(k))(K_s+S(k))}$$

$$a_{23} = \frac{k_{smax}TsX_v(k)Gs\,e^{Gs(S(k)-Ps)}}{\left(1+e^{Gs(S(k)-Ps)}\right)^2} - \frac{TsX_v(k)S(k)DO(k)\mu_{max}}{\left(K_d+DO(k)\right)\left(K_s+S(k)\right)^2} + \frac{TsX_v(k)DO(k)\mu_{max}}{\left(K_d+DO(k)\right)(K_s+S(k))}$$

$$a_{24} = -\frac{TsX_v(k)S(k)DO(k)\mu_{max}}{\left(K_d+DO(k)\right)^2\left(K_s+S(k)\right)} + -\frac{TsX_v(k)S(k)\mu_{max}}{\left(K_d+DO(k)\right)\left(K_s+S(k)\right)}$$

$$a_{31} = 0$$

$$a_{32} = -\frac{Ts\,S(k)\,DO(k)\mu_{max}}{Y_{x/s}\left(K_d + DO(k)\right)\left(K_s + S(k)\right)} + Ts\,m_s$$

$$a_{33} = 1 + \frac{Ts\,X_v(k)S(k)DO(k)\mu_{max}}{Y_{x/s}\left(K_d + DO(k)\right)\left(K_s + S(k)\right)^2} - \frac{Ts\,X_v(k)DO(k)\mu_{max}}{Y_{x/s}\left(K_d + DO(k)\right)\left(K_s + S(k)\right)}$$

$$a_{34} = \frac{Ts\,X_v(k)S(k)DO(k)\mu_{max}}{Y_{x/s}\left(K_d + DO(k)\right)^2\left(K_s + S(k)\right)} - \frac{Ts\,X_v(k)S(k)\mu_{max}}{Y_{x/s}\left(K_d + DO(k)\right)\left(K_s + S(k)\right)}$$

$$a_{41} = K_2\,Ts$$

$$a_{42} = K_2\,Ts - \frac{k_{emax}\,Ts\,S(k)\,DO(k)u_{max}}{(1 + e^{Ge(Pe-Tsk)})(K_d + DO(k))(K_s + S(k))} + \frac{k_{emax}\,Ts\,S(k)\,DO(k)u_{max}}{(1 + e^{Ge(Pe-Tsk)})(K_d + DO(k))(K_s + S(k))}$$

$$a_{43} = \frac{k_{emax}\,Ts\,X_v(k)S(k)\,DO(k)u_{max}}{(1 + e^{Ge(Pe-Tsk)})(K_d + DO(k))(K_s + S(k))^2} - \frac{k_{emax}\,Ts\,X_v(k)S(k)\,DO(k)u_{max}}{(1 + e^{Ge(Pe-Tsk)})(K_d + DO(k))(K_s + S(k))^2} +$$

$$\frac{k_{emax}\,Ts\,X_v(k)\,DO(k)u_{max}}{(1 + e^{Ge(Pe-Tsk)})(K_d + DO(k))(K_s + S(k))} - \frac{k_{emax}\,Ts\,X_v(k)\,DO(k)u_{max}}{(1 + e^{Ge(Pe-Tsk)})(K_d + DO(k))(K_s + S(k))}$$

$$a_{44} = 1 - K_3 Q_A\,Ts + \frac{k_{emax}\,Ts\,X_v(k)S(k)\,DO(k)u_{max}}{(1 + e^{Ge(Pe-Tsk)})(K_d + DO(k))^2(K_s + S(k))} - \frac{k_{emax}\,Ts\,X_v(k)S(k)\,DO(k)u_{max}}{(1 + e^{Ge(Pe-Tsk)})(K_d + DO(k))^2(K_s + S(k))} +$$

$$+ \frac{k_{emax}\,Ts\,X_v(k)S(k)u_{max}}{(1 + e^{Ge(Pe-Tsk)})(K_d + DO(k))(K_s + S(k))} - \frac{k_{emax}\,Ts\,X_v(k)S(k)u_{max}}{(1 + e^{Ge(Pe-Tsk)})(K_d + DO(k))(K_s + S(k))}$$

| **Predict Step** |
| --- |
| $\hat{x}(k\|k-1) = f(\hat{x}(k-1\|k-1), k)$ |
| $P(k\|k-1) = A(k)P(k\|k-1)A^T(k) + Q$ |
| **Update Step** |
| $\tilde{y}(k) = y(k) - H\hat{x}(k\|k-1)$ |
| $K(k) = P(k\|k-1)H^T\left[HP(k\|k-1)H^T + R\right]^{-1}$ |
| $\hat{x}(k\|k) = \hat{x}(k\|k-1) + K(k)\tilde{y}(k)$ |
| $P(k\|k) = \left(I + K(k)H\right)P(k\|k-1)$ |

Table 3. EKF algorithm

## Appendix B. Decentralized Kalman Filter

Decentralized Filtering (no feedback to the local filters)

i. Information matrix update

$$P^{-1} = \left(P^{-}\right)^{-1} + H^{T} \quad R^{-1} \quad H \tag{B1}$$

ii. Gain computation

$$K = P\,H^{T}R^{-1} \tag{B2}$$

iii. Estimate update

$$\hat{x} = \hat{x}^{-} + K \quad \left(x - H \quad \hat{x}^{-}\right) \tag{B3}$$

iv. Project ahead to next step

$$\hat{x}^{-}_{k+1} = \varphi_k \quad \hat{x}_k \tag{B4}$$

$$P^{-}_{k+1} = \varphi_k \quad P_k \quad \varphi_k^{T} + Q_k \tag{B5}$$

Recall that $\mathbf{P}^{-1}$ is called the *information matrix*. In terms of information, i) says that the updated information is equal to the prior information plus the additional information obtained from the measurement at time $t_k$.. Furthermore, if $\mathbf{R}_k$ is block diagonal, the total "added" information can be divided into separate components, each representing the contribution from the respective measurement blocks. That is, we have (omitting the k subscripts for convenience).

$$H^{T}R^{-1}H = H_1^{T}R_1^{-1}H_1 + H_2^{T}R_2^{-1}H_2 + \ldots + H_N^{T}R_N^{-1}H_N \tag{B6}$$

We also note that the estimate update equation al time $t_k$ can be written in a different form as follows:

$$\begin{aligned}
\hat{x} &= \left(I - K \quad H\right) \quad \hat{x}^{-} + K \quad X \\
\hat{x} &= \left(I - P \quad H^{T} \quad R^{-1} \quad H\right) \quad \hat{x}^{-} + P \quad H^{T} \quad R^{-1} \quad X \\
\hat{x} &= \left(P \quad P^{-1} - P \quad H^{T} \quad R^{-1} \quad H\right) \quad \hat{x}^{-} + P \quad H^{T} \quad R^{-1} \quad X \\
\hat{x} &= P \quad \left[\left(P^{-1} - H^{T} \quad R^{-1} \quad H\right) \quad \hat{x}^{-} + H^{T} \quad R^{-1} \quad X\right] \\
\hat{x} &= P \quad \left[\left(P^{-}\right)^{-1} \quad \hat{x}^{-} + H^{T} \quad R^{-1} \quad X\right]
\end{aligned} \tag{B7}$$

When writing in this form, it is clear that the updated estimate is a linear blend of the old information with the new information.

For simplicity, we will start with just two local filters in our decentralized system, and we will continue to omit the k subscripts to save writing. Both filters are assumed to implement the full-order state vector, and at step k both are assumed to have available their respective prior estimates $\mathbf{m_1}$ and $\mathbf{m_2}$ and their associated error covariances $\mathbf{M_1}$ and $\mathbf{M_2}$. For Gaussian Process, m1 and m2 will be the means of x conditioned on their respective measurement streams up to, but not including, time $t_k$. The measurements presented to filters 1 and 2 at time $t_k$ are $X_1$ and $X_2$ and they have the usual relationship to x:

$$X_1 = H_1 x + v_1 \tag{B8}$$

$$X_2 = H_2 x + v_2 \tag{B9}$$

Where $\mathbf{v_1}$ and $\mathbf{v_2}$ are zero mean random variables with covariances $\mathbf{R_1}$ and $\mathbf{R_2}$. The estate $\mathbf{x}$ and noises $\mathbf{v_1}$ and $\mathbf{v_2}$ are assumed to be mutually uncorrelated as usual.
If we assume now that local filters 1 and 2 do not have access to each other's measurements, the filters will form their respective error covariances and estimates according to (B1) and (B7).

**Local filter 1**

$$P_1^{-1} = M_1^{-1} + H_1^T \; R_1^{-1} \; H_1 \tag{B10}$$

$$\hat{X}_{1OPT} = P_1 \left( M_1^{-1} \; m_1 + H_1^T \; R_1^{-1} \; X_1 \right) \tag{B11}$$

**Local filter 2**

$$P_2^{-1} = M_2^{-1} + H_2^T \; R_2^{-1} \; H_2 \tag{B12}$$

$$\hat{X}_{2OPT} = P_2 \left( M_2^{-1} \; m_2 + H_2^T \; R_2^{-1} \; X_2 \right) \tag{B13}$$

Note that the local estimates will be optimal, conditioned on their respective measurement streams, but not with respect to the combined measurements.
Now consider the master filter. It is looking for an optimal global estimate of $\mathbf{x}$ conditioned on both measurement streams 1 and 2.
$\mathbf{m}$ = optimal estimate of $\mathbf{x}$ conditioned on both measurement streams up to but not including $t_k$
$\mathbf{M}$ = covariance matrix associated with $\mathbf{m}$
The optimal global estimate and associated error covariance are then

$$P^{-1} = \begin{bmatrix} H_1^T & H_2^T \end{bmatrix} \begin{bmatrix} R_1^{-1} & 0 \\ 0 & R_2^{-1} \end{bmatrix} \begin{bmatrix} H_1 \\ H_2 \end{bmatrix} + M^{-1} \tag{B14}$$

$$P^{-1} = M^{-1} + H_1^T \; R_1^{-1} \; H_1 + H_2^T \; R_2^{-1} \; H_2$$

$$\hat{x} = P \left( M_{-1} \; m + H_1^T \; R_1^{-1} \; H_1 + H_2^T \; R_2^{-1} \; H_2 \right) \tag{B15}$$

However, the master filter does not have direct access to **X1** and **X2**, so we will rewrite (B14) and (B15) in terms of the local filter's computed estimates and covariances. The result is:

$$P^{-1} = \left(P_1^{-1} - M_1^{-1}\right) + \left(P_2^{-1} - M_2^{-1}\right) + M^{-1} \tag{B16}$$

$$\hat{x} = P\left[\left(P_1^{-1}\hat{x}_1 - M_1^{-1}m_1\right) + \left(P_2^{-1}\hat{x}_2 - M_2^{-1}m_2\right) + M^{-1}m\right] \tag{B17}$$

It can now be seen that the local filters can pass their respective $\hat{x}_i$, $P_i^{-1}$, $m_i$ and $M_i^{-1}$ (i=1,2) on to the master filter, which, in turn, can then compute its global estimate. The local filters can, of course, do their own local projections and then repeat the cycle at step (k+1). Likewise, the master filter can project its global estimate and get a new **m** and **M** for the next step. Thus, it can be seen that this architecture permits complete autonomy of the local filters, and it yields local optimality with respect to the respective measurement stream. The system also achieves global optimality in the master filter.

## Appendix C: Bayesian Regression framework and Gaussian Process Regression

Suppose that we have a noisy training data set *D* which consists of *m* pairs of *n*-dimensional input vectors $\{x_i\}$ (regression vector) joined in a $n \times m$ matrix *X*, and *m* scalar noisy observed outputs $\{y_i\}$ collected in a vector $\boldsymbol{y}$.

$$D = \left\{\left(x_i, y_i\right) \middle| i = 1, \ldots, m\right\} = \{X, y\} \tag{C1}$$

In order to construct a probabilistic statistical model for *D*, the following data-generating process is assumed:

$$y_i = f\left(x_i\right) + \varepsilon_i \tag{C2}$$

where the latent real-valued function *f* is the deterministic or systematic component of the model, and the additive random term $\varepsilon$ is the observation error. The aim of regression is to identify the systematic component *f* from the empirical observations *D*. The Bayes' rule (C3) shows the components of Bayesian Inference (Bernardo & Smith, 2006): the joint likelihood, the prior distribution, and the posterior distribution. Bayesian inference alludes to the process of updating our beliefs according to Bayes' rule, i.e. computing the posterior from likelihood and the prior, integrating the information contained in the observed data.

$$\overbrace{p\left(f \middle| D, \theta_L, \theta_P\right)}^{\text{Posterior}} = \frac{\overbrace{p\left(y \middle| f, \theta_L\right)}^{\text{Likelihood}}\overbrace{p\left(f \middle| X, \theta_P\right)}^{\text{Prior}}}{\underbrace{p\left(D \middle| \theta_P, \theta_L\right)}_{\text{Evidence}}} \tag{C3}$$

Where $P(D|.) = P(y|X,.)$, $P(.|D,.) = P(.|y,X,.)$, $y_i = y(x_i)$ $y = [y_1, L, y_m]^T$, denotes the observed outputs, $f_i = f(x_i)$ $f = [f_1, L, f_m]^T$, are the latent function values, and $\theta_L$, $\theta_P$ denote additional parameters (hyperparameters) of the likelihood and prior distribution respectively.

The evidence or marginal likelihood is the normalising constant appearing in the denominator of Bayes' rule. This quantity is one of the most useful quantities in the Bayesian framework (e.g., in hypothesis testing applications) (O'Hagan, 2004; Bernardo & Smith, 2006), however the evidence is not considered in the remainder of the exposition.

If the measurements of the training data set *D* are independent, and the observation error $\varepsilon$ it is assumed that is normal, independent and identically distributed with mean zero and variance $\sigma^2$, then, in this case the joint likelihood is:

$$p(y|f,\sigma^2) = \prod_{i=1}^{m} N(f_i,\sigma^2) = N(f,\sigma^2 I_{m\times m}) \tag{C4}$$

In the Bayesian non-parametric approach, a prior is put directly on the space of functions and the inference is carried out on *f*. The prior distribution is usually chosen from a parametric family of distributions or a mixture of these. The expression "Gaussian Process Regression model" refers to using a Gaussian Process as a prior on **f**. This means that every finite-dimensional marginal joint distributions of function values **f** associated to any input subset of **X** is multivariate Gaussian.

$$p\left(f|X,\theta_P\right) = N\left(m\left(X\right),K\left(X,\theta_P\right)\right) \tag{C5}$$

A Gaussian Process is fully specified by a mean function $m(X) = [m(x_1), L, m(x_m)]^T$ and a positive-definite covariance matrix $K(X, \theta_P)$, and it can be viewed as a generalization of the multivariate Gaussian distribution to infinite dimensional objects. Choosing a particular form of covariance function, the hyperparameters $\theta_P$ may be introduced to the Gaussian Process prior. Depending on the actual form of the covariance function $K(X, \theta_P)$ the hyperparameters $\theta_P$ can control various aspects of the Gaussian Process. For simplicity the prior mean function is set to be zero $\mu(X)=0$, this is completely general provided that a constant term is included in the covariance function (Williams & Rasmussen, 1996; Kuss, 2006).

The posterior distribution over function values is obtained introducing (C4) and (C5) in (C3)

$$p\left(f|D,\sigma^2,K\right) \propto N\left(f,\sigma^2 I_{m\times m}\right) N\left(0,K\right) \propto N\left(K\left(K+\sigma^2 I_{m\times m}\right)^{-1} y,\left(K^{-1}+\sigma^{-2} I_{m\times m}\right)^{-1}\right) \tag{C6}$$

The distribution of the latent function value $f_* = f(x_*)$ for an arbitrary new input $x_*$ conditioned on the training function outputs is (Kuss, 2006):

$$p\left(f_*\middle|f, x_*, X, K\right) \propto N\left(k_*^T K^{-1} f, k_{**} - k_*^T K^{-1} k_*\right) \tag{C7}$$

where $k_* = [K(x_*, x_1), L, K(x_*, x_m)]^T$ is a vector of prior covariances between $x_*$ and the training inputs $\mathbf{X}$, and $k_{**} = K(x_*, x_1)$.

The predictive distribution of $f_*$ is obtained by integrating out the training function values $\mathbf{f}$ from (B6) over the posterior distribution (C7). The predictive distribution is again multivariate normal:

$$p\left(f_*\middle|D, x_*, K, \sigma^2\right) = \int p\left(f_*\middle|f, x_*, X, K\right) p\left(f\middle|D, \sigma^2, K\right) df \propto N\left(k_*^T C^{-1} y, k_{**} - k_*^T C^{-1} k_*\right) = N(\hat{f}_*, \sigma^2_{\hat{f}_*}) \tag{C8}$$

The predictive uncertainty, i.e. the covariance matrix of $f_*$, does not depend on $y$, but only on the dependencies induced by the covariance as a function of $x_*$ and $\mathbf{X}$. This can be generalized to an arbitrary set of new inputs $\mathbf{X}_*$, meaning that the posterior process $f \mid D$ is again a Gaussian Process with posterior mean and covariance function

$$\begin{aligned} \hat{f}_* &= E\left(f_*\middle|D, x_*, K, \sigma^2\right) = k_*^T C^{-1} y \\ \sigma^2_{\hat{f}_*} &= k_{**} - k_*^T C^{-1} k_* \end{aligned} \tag{C9}$$

where $C = C(X, \theta_p, \sigma^2) = K(X, \theta_p) + \sigma^2 I_{m\times m}$.

The resulting posterior (C6) and the predictive distribution (C8) are of the same family of distributions as the prior (C5). The class of prior distributions with this property is called conjugate to a likelihood model (O'Hagan, 2004; Bernardo & Smith, 2006). The calculations are analytically tractable only for conjugate models with normal noise. For all other models the posterior and the predictive distribution cannot be computed analytically, so techniques for approximate inference have been used, for example, Markov Chain Monte Carlo (MCMC) sampling techniques (Williams & Rasmussen, 1996; Neal, 1997; Kuss, 2006).

**Covariance Function**

The elements of the parameterized covariance matrix, $C(X, \theta_p, \sigma^2)$, are denoted $Cij = C(x_i, x_j)$, and they are functions of the training input data $\mathbf{X}$, because these data determine the correlation between the training data outputs $y$. A suitable parametric form of the covariance function is:

$$Cij = \theta_0 + \theta_1 \exp\left[-\frac{1}{2}\sum_{l=1}^{n} \frac{\left(x_i^{(l)} - x_j^{(l)}\right)^2}{r_l^2}\right] + \theta_2\,\delta(i, j) + \sum_{l=1}^{n} \alpha_l\, x_i^{(l)} x_j^{(l)} \tag{C10}$$

where $x_i^{(l)}$ is the $l^{th}$ dimension of the input vector, $x_i$.

The four terms in this equation are now briefly described.

i- A bias term $\theta_0$ controlling the scale of the bias contribution to the covariance. The constant term $\theta_0$, adds a constant offset to the estimated latent function value $f_* = f(x_*)$. This justifies assigning $m(X) = 0$ to the prior mean function (C5) without loss generality.

ii- The exponential term (involving $\theta_1$ and *rl*) expresses our belief that inputs which are close to each other give rise to outputs which are close to each other or that are highly correlated; the *rl* hyperparameters allow a different distance measure for each input dimension, and $\theta_1$ gives the overall scale of variations in the output space (local correlations).

iii- The third term involving the hyperparameter $\theta_2 = \sigma^2$, i.e., the variance of the noise model for the outputs and therefore only occurs in $Cij$ when $i = j$.

iv- The fourth and last term characterizes the nonstationarity of the covariance function. It is a linear regression term and involve $\alpha_1$, $l = 1, L, n$, these hyperparameters controlling the scale of the linear trends to the covariance.

Besides (C10), there are other forms of the covariance function which could be used. The only restriction is that the covariance matrix be positive definite. Abrahamsen has written a comprehensive survey on numerous valid covariance functions (Abrahamsen, 1997).

**Hyperparameters Determination**

The hyperparameters of the covariance function are not known in advance, and they must be determined using the training data. The literature has reported several approaches to hyperparameter estimation: Cross-Validation method (Wahba, 1990), Evidence Maximization (MacKay, 1992; Gibbs, 1997), Monte Carlo methods (Neal, 1997), Maximum Likelihood, and Maximum a Posteriori method (Rasmussen, 1996). In this work the last method is used, i.e., the Maximum a Posteriori (MAP) approach.

For implementation purpose, the hyperparameters vector is defined as:

$$\log\theta = [\log\theta_0, \log\theta_1, \log r_1, \ldots, \log r_n, \log\theta_2, \log\alpha_1, \ldots, \log\alpha_n]^T$$

The number of hyperparameters of the covariance function (C10) increases linearly with $n$, the dimension of the input space, i.e., $\#\theta = 2n + 3$. The likelihood of the parameters is $L(\theta) = p(D \mid C(\theta)) = p(D \mid \theta, C(.))$, where $C(.)$ specifies the form of covariance function. The maximum likelihood approach is to maximize $L(\theta)$ to yield the optimum hyperparameters. An improvement over this is to incorporate a prior, $p(\theta)$, on the hyperparameters. By Bayes theorem, the posterior probability of the hyperparameters, given the training data, is:

$$p(\theta \mid D, C(.)) = \frac{p(D \mid \theta, C(.))p(\theta \mid X, C(.))}{p(D \mid C(.))} \qquad \text{(C11)}$$

Maximization of $p(\theta \mid D, C(.))$ is known as the MAP approach, which is a Bayesian version of maximum likelihood estimation. It is possible to analytically express the posterior $p(\theta \mid D, C(.))$ and its partial derivatives with respect to hyperparameters $\{\theta i\}$ as derived, for example, in Mardia & Marshall (1984). Let $L=\log\, p(\theta \mid D, C(.))$, then:

$$\begin{cases} L = -\frac{1}{2}\log|C| - \frac{1}{2}y^T C^{-1} y - \frac{m}{2}\log 2\pi + \log p(\theta) + c \\ \frac{\partial L}{\partial \theta_i} = -\frac{1}{2}\mathrm{trace}\left(C^{-1}\frac{\partial C}{\partial \theta_i}\right) + \frac{1}{2}y^T C^{-1}\frac{\partial C}{\partial \theta_i}C^{-1}y + \frac{\partial \log p(\theta)}{\partial \theta_i} \end{cases} \quad (C12)$$

where the prior on $\boldsymbol{\theta}$, $p(\theta)$ is assumed to be independent of the other priors.
The MAP estimates are found using (C12) in a gradient descent, or conjugate gradient optimization techniques to locate a local maximum of the posterior distribution. This approach has the advantage that a reasonable approximation to a local maximum can be found with relatively few function and gradient evaluations. On the other hand, the posterior distribution is often multimodal, and the risk of using gradient optimization techniques is that the algorithm can keep trapped in bad local maxima. In order to reduce this problem, suitable priors can be assigned and multiple random restarts for the optimization routines can be fulfilled. The initialization of the hyperparameters is important, because improper initial values will make the partial derivatives of the likelihood very small, thus creating problems for the optimization algorithm.
In order to implement the algorithm (equations (C9), (C10), and (C12)) it was necessary to invert the covariance matrix $\mathbf{C}$. Any exact inversion method has an associated computational cost that is $O(m^3)$, moreover, direct inversion implementation can run into numerical problems because $\mathbf{C}$ is generally ill-conditioned, i.e., the condition number is large. In order to improve the condition number of the matrix inversion operation, $C^{-1}$ can be computed indirectly by using Cholesky, LU, or SVD decomposition (note that the positive definiteness property of the covariance matrix is guaranteed). All these methods also require $O(m^3)$ operations (Golub & Van Loan, 1990).
It is important to observe that the optimization routines require in each step the evaluation of the gradient of the log likelihood, i.e., the computation of $C^{-1}$ and so calculating gradients becomes time consuming for large training data sets. If $m$ is large ($m > 10^3$), Skilling's approximate inversion methods, which are $O(m^2)$, can be used (Skilling, 1993).

## 5. References

Abrahamsen, P. (1997). A review of Gaussian random fields and correlation functions. *Technical Report 917*, Norwegian Computing Center, Oslo, Norway, 2nd edition.

Amicarelli A. Modelado, Identificación y Control de Bioprocesos. (2009). (In Spanish) *PhD thesis.* Universidad Nacional de San Juan Argentina - Instituto de Automática. ISBN978-987-05-7087-5, pp. 1-191.

Amicarelli, A.; di Sciascio, F.; Álvarez, H. & Ortiz, O. (2006). (In Spanish) Estimación de Biomasa en un proceso Batch: Aplicación a la producción de δ-endotoxinas de Bt. In: *XXII Interamerican congress of chemical engineering*.

Amicarelli, A., di Sciascio F., Toibero J. M. & Álvarez, H. (2010). Including dissolved oxygen dynamics to the Bacillus thuringiensis δ -endotoxins production process. *Brazilian Journal of Chemical Engineering*, Vol. 27, No. 01, pp. 41 - 62.

Anderson, B.D.O., & Moore, J.B. (1979). Optimal Filtering. Prentice-Hall, Englewood Cliffs, New Jersey. ISBN 0136381227 : 0136381227.

Aronson, A. I. (1993). The two faces of Bacillus thuringiensis: insecticidal proteins and post exponential survival. *Molecular Microbiology*, Vol. 7, pp. 489–496.

Atehortúa, P., Álvarez, H., & Orduz, S. (2006). Comments on: A sporulation kinetic model for batch growth of B. thuringiensis. *The Canadian Journal of Chemical Engineering*, Vol. 84, No. 03, pp. 386-388.

Atehortúa, P., Alvarez, H., & Orduz, S. (2007). Modeling of growth and sporulation of Bacillus thuringiensis in an intermittent fed batchculturewith total cell retention. *Bioprocess and Biosystems Engineering*, Vol. 30, pp. 447–456.

Bastin, G. & Dochain, D. (1990). On-line estimation and adaptive control of bioreactors. Amsterdam: Elsevier. ISBN-13: 978-0444884305.

Bernardo, J. M., & Smith, A. F. (2006). Bayesian Theory (2nd ed.) Chichester: Wiley.

Bishop C.M. (1995, reprint 2005). Neural Networks for Pattern Recognition, Oxford University Press.

Brawn R. , Hwang P. (1997) .Introduction to Random Signals and Applied Kalman Filtering. 3rd ed. John Wiley & Sons: New York, pp. 371-375.

Daum, F. (2005). Nonlinear Filters: Beyond the Kalman Filter. IEEE *Aerospace and Electronic Systems Magazine,* Vol. 20, No 8, Part 2, pp. 57- 69.

di Sciascio, F. & Amicarelli, A. (2008). Biomass Estimation in Batch Biotechnological Processes by Bayesian Gaussian Process Regression, *Computers and Chemical Engineering* Vol. 32, pp. 3264 - 3273.

Dochain, D. (2003). State and Parameter Estimation in Chemical and Biochemical Processes: A Tutorial. *Journal of Process Control*, Vol. 13, No. 8, pp. 801–818.

Ghribi, D., Zouari, N., Trabelsi, H., & Jaoua, S. (2007). Improvement of Bacillus thuringiensis delta-endotoxin production by overcome of carbon catabolite repression through adequate control of aeration. *Enzyme and Microbial Technology*, Vol. 40, No. 4, pp. 614–622.

Gibbs, M.N. (1997). Bayesian Gaussian Processes for Regression and Classification, PhD thesis, University of Cambridge.

Golub, G.H., Van Loan, C.F. (1990). Matrix Computation, Baltimore, John Hopkins University Press.

Hastie, T., Tibshirani, R., & Friedman, J.H. (2001). The Elements of Statistical Learning. Springer Verlag.

Haykin, S. (1999). Neural Networks: A comprehensive Foundation. Second Edition by Prentice-Hall, Inc., New Jersey.

Jazwinski, A.H. (1970). Stochastic Processes and Filtering Theory, Academic Press, Inc. New York.

Kuss, M. (2006). Gaussian Process Models for Robust Regression, Classification, and Reinforcement Learning. *PhD thesis*, Technische Universität Darmstadt.

Leal Ascencio, R. (2001). Artificial neural networks as a biomass virtual sensor for a batch process. *In Proceedings of the 2001 IEEE interational symposium on intelligent control.*

Li, B. (2003). Artificial neural network based software sensor for biomass during microorganism cultivation. *PhD thesis.* South China University of Technology.

Lind, I. (2006). Regressor and Structure Selection Uses of ANOVA in System Identification. *PhD thesis,* Department of Electrical Engineering, University of Linköping, Sweden.

Little, R., & Rubin, D. (2002). Statistical Analysis with Missing Data. New York, Wiley.

Liu, B. L., & Tzeng, Y. M. (2000). Caracterization study of the sporulation kinetics of Bacillus thuringiensis. *Biotechnology and Bioengineering,* Vol. 68, No 1, pp. 11–17.

Ljung, L. (2006). Some Aspects on Nonlinear System Identification. *14th IFAC Symposium on System Identification,* SYSID-2006, Newcastle, Australia.

MacKay, D.J. (1992). Bayesian interpolation. Neural Computation 4-33, pp. 415–447.

Mardia, K.V., & Marshall, R.J. (1984). Maximum Likelihood estimation of models for residual covariance in spatial regression. *Biometrika,* Vol. 71, No 1, pp. 135-146.

Mannale, R. (2006). Comparison of Regressor Selection Methods in System Identification. PhD thesis, Department of Electrical Engineering, University of Linköping, Sweden.

Narendra, K., & Parthasarathy, K. (1990). Identification and Control of Dynamical Systems Using Neural Networks. *IEEE Transaction On Neural Networks.*

Neal, R. M. (1996). Bayesian learning for neural networks. Springer-Verlag New York. ISBN 0-387-94724-8.

Neal, R. M. (1997). Monte Carlo implementation of GP models for Bayesian regression and classification. *Tech. Rep.* No. 9702. Toronto: Department of Statistics, University of Toronto.

O'Hagan, A. (2004). Bayesian Inference (2nd ed.). Kendall's Advanced Theory of Statistics 2B. London: Edward Arnold.

Rasmussen, C.E. (1996).Evaluation of Gaussian Processes and Other Methods for Nonlinear Regression. *PhD thesis* University of Toronto.

Rivera, D., Margaritis, A., & De Lasa, H. (1999). A sporulation kinetic model for batch growth of B. thuringiensis. *Canadian Journal of Chemical Engineering,* Vol. 77, pp. 903–910.

Simon, D. (2006), Optimal State Estimation: Kalman, H∞, and Nonlinear Approaches. John Wiley & Sons, Inc., Hoboken, New Jersey.

Skilling, J. (1993) Bayesian numerical analysis, In Physics and Probability, ed. by W. Grandy and P. Milonni, Cambridge University Press.

Starzak, M. & Bajpai, R. (1991). A structured model for vegetative growth and sporulation in Bacillus thuringiensis. *Applied Biochemistry and Biotechnology,* Vol. 28/29, pp. 699–718.

Vallejo, F., González, A., Posada, A., Restrepo, A., & Orduz, S. (1999). Production of Bacillus thuringiensis subsp. Medellín by batch and fed-batch culture. *Biotechnology Techniques*, Vol. 13, pp. 279–281.

Williams, K. I,. & Rasmussen, C. E. (1996) Gaussian Processes for Regression. MIT Press.

Wahba, G. (1990) Spline Models for Observational Data. Volume 59 of Series in Applied Mathematics. SIAM, Philadelphia.

# 5

# Current Status of Woody Biomass Utilization in ASEAN Countries

Takahiro Yoshida[1,*] and Hidenori Suzuki[2]
[1]Department of Wood Processing, Forestry and Forest Products Research Institute (FFPRI)
[2]Department of Forest Engineering, Forestry and Forest Products Research Institute (FFPRI)
[1,2]Matsu-no-sato, Tsukuba, Ibaraki
Japan

## 1. Introduction

Renewable energy, including biomass, has received increasing attention because of worldwide efforts to prevent global warming and alleviate soaring oil prices. When biomass is used as an energy source, for example, it is converted to ethanol as an alternative to gasoline or burned in a boiler to generate heat and power. Cereals such as corn are promising candidates for easily convertible biomass for ethanol production. However, it is possible that the supply of such potential biofuels will become unstable because of conflicts with food production. For example, corn prices have doubled or have reached levels not seen in many years[11]. In contrast, woody biomass is inedible and thus may be a promising candidate as a future renewable energy source. Woody biomass can be collected in large amounts from forests or as a by-product of the forest industry. However, forests are unevenly distributed throughout the world, and so the distribution of woody biomass is uneven. Countries belonging to the Association of Southeast Asian Nations (ASEAN) have a vast range and abundance of forest resources. However, these resources have been considerably reduced because of unrestrained logging in some countries. Sustainable forest management is necessary for the continued use of the available biomass as a renewable energy source.

Our objective was to investigate the current status of the use and consumption of woody biomass to discuss future possibilities and difficulties in the use of woody biomass as an energy source. We focused on Cambodia, Indonesia, Laos, Malaysia, the Philippines, Thailand, and Vietnam. Hereafter, our use of the term "ASEAN countries," unless otherwise specified, refers to these seven countries.

## 2. Forests, forestry, and biomass

Table 1 is an overview of the biomass, including aboveground, belowground, and deadwood biomass, and growing stock in the forests of the ASEAN countries. Indonesia has the most extensive forest area and biomass stock, but the growing stock has been decreasing since 1990, probably because of illegal logging or forest fires. Malaysia has the highest biomass and growing stock by area, whereas Thailand has the lowest.

The data for the main forest products differs among the ASEAN countries because they were obtained from different sources. Nevertheless, we conducted our investigation using this data because it was uniform within each country (Table 2). Indonesia and Malaysia produce substantial amounts of sawnwood and plywood, whereas Thailand is active in the production of wood chips and particles. In Vietnam, the production of sawnwood is active and very high compared to that of industrial roundwood. Cambodia and Laos consume a large proportion of roundwood as fuel.

| | Cambodia | Indonesia | Laos | Malaysia | Philippines | Thailand | Vietnam |
|---|---|---|---|---|---|---|---|
| Forest area (Kha) | 10,447 | 88,495 | 16,142 | 20,890 | 7,162 | 14,520 | 12,931 |
| Biomass stock (Mt) | 2,811 | 13,090 | 3,301 | 8,073 | 2,156 | 1,592 | 2,606 |
| Biomass stock by area (t/ha) | 269 | 148 | 204 | 386 | 301 | 110 | 202 |
| Growing stock ($Mm^3$) | 998 | 5,216 | 957 | 5,242 | 1,248 | 599 | 850 |
| Growing stock by area ($m^3$/ha) | 96 | 59 | 59 | 251 | 174 | 41 | 66 |
| Growing stock by area ($m^3$/ha) | 96 | 59 | 59 | 251 | 174 | 41 | 66 |
| Annual change in growing stock | | | | | | | |
| 1990-2000 ($m^3$/y/ha) | -0.11 | -3.33 | 0.00 | 1.94 | 0.13 | 0.00 | -0.26 |
| 2000-2005 ($m^3$/y/ha) | -0.11 | -4.61 | 0.00 | 1.94 | 0.08 | 0.00 | -0.40 |

Table 1. Biomass* and growing stock in a forest

*Biomass includes above- and below-ground biomass, dead-wood biomass.

[5, 19]: Reference source modified by the authors

| | Cambodia | Indonesia | Laos | Malaysia | Philippines | Thailand | Vietnam |
|---|---|---|---|---|---|---|---|
| Industrial Roundwood ($Km^3$) | 118 | 35,551 | 194 | 25,169 | 2,869 | 8,700 | 5,850 |
| Sawnwood ($Km^3$) | 4 | 4,330 | 130 | 5,142 | 358 | 2,868 | 5,000 |
| Plywood ($Km^3$) | 5 | 3,353 | 24 | 5,601 | 235 | 120 | 70 |
| Chips and Particles ($Km^3$) | - | 1,100 | - | 501 | - | 2,080 | 2,500 |
| Fiberboard ($Km^3$) | - | 427 | - | 2,633 | 0 | 883 | 180 |
| Particle Board ($Km^3$) | - | 297 | 0 | 222 | 6 | 2,600 | 48 |
| Wood Residues ($Km^3$) | - | 388 | - | - | - | - | 350 |
| Woodfuel ($Km^3$) | 8,735 | 73,720 | 5,944 | 2,908 | 12,581 | 19,503 | 22,000 |
| Wood Charcoal (Kt) | 34 | 681 | 20 | 54 | 610 | 1,326 | 109 |
| Total (Roundwood) ($Km^3$) | 8,853 | 109,270 | 6,137 | 28,077 | 15,450 | 28,203 | 27,850 |

Table 2. The production of forest products in ASEAN countries (2008)

- : Insufficient data

[4, 19]: Reference source modified by the authors. The sum of wood product (= sawnwood + plywood + chips and particles + fiberboard + particle board + wood residues) does not coincide with the amount of industrial roundwood.

## 3. Energy consumption

In terms of energy supply and consumption in the ASEAN countries (Table 3), Indonesia has the largest total primary energy supply (TPES) at 191 million tons of oil equivalent (Mtoe), whereas Malaysia has the largest per capita TPES at 2.73 toe. The bulletin of the ASEAN Center for Energy (ACE) states that the energy consumption was approximately 1.41 Mtoe in Laos in 1999. The energy consumption in Laos was 0.27 toe per capita, based on population data reported by the National Statistics Center of the Lao PDR[16] in 2000. The consumption of renewable energy accounts for 2% of the TPES in Malaysia and 46% in Vietnam. Cambodia consumed 3.41 Mtoe of renewable energy from primary solid biomass, which accounts for 66% of the energy consumed in 2007. Laos consumes approximately 66% of its total energy from fuelwood, charcoal, and sawdust. However, the difficulties encountered in comparing these values are that the data source and the year the data was collected are different for each country, and Malaysia's dependency on biomass energy is insignificant, whereas Cambodia, Laos, and Vietnam depend heavily on these sources of energy. This is partly because few households have electricity; for example, in 2004, only 47% of households in Laos[15] and only 12% in Cambodia[14] had electricity.

| | Cambodia | Indonesia | Laos | Malaysia | Philippines | Thailand | Vietnam |
|---|---|---|---|---|---|---|---|
| TPES (Mtoe) | 5.13 [9] | 190.64 [9] | | 72.59 [9] | 39.98 [9] | 103.99 [9] | 55.79 [9] |
| TPES per population (toe/capita) | 0.36 [9] | 0.84 [9] | | 2.73 [9] | 0.45 [9] | 1.63 [9] | 0.66 [9] |
| Consumption of renewable energy from primary solid biomass* (Mtoe) (Proportion to TPES) | 3.41 [9] (66%) | 50.50 [9] (26%) | | 1.38 [9] (2%) | 4.87 [9] (11%) | 8.79 [9] (9%) | 23.28 [9] (46%) |
| Energy consumption (Mtoe) | | | 1.41 [12] | | | | |
| Energy consumption per population (toe/capita) | | | 0.27 [12, 16] | | | | |
| Data year | 2007 | 2007 | 1999 | 2007 | 2007 | 2007 | 2007 |

Table 3. Energy supply and consumption

TPES: Total Primary Energy Supply = Indigenous production + imports - exports - international marine bunkers ± stock changes
* Data are also available for charcoal.
[9,12,14,16,19]: Reference source modified by the authors.

## 4. Woody biomass use

### 4.1 Logging residues

We investigated the use of logging residues in Cambodia[18, 24], Indonesia[17], and Malaysia.[6, 7, 13]. Although rubber plantations are generally expected to regenerate every 30–35 years, the tapping cycle of para rubber trees (PRT; *Hevea brasiliensis*) in Cambodia[18] is longer because of interruptions due to civil war. Woody residues that remain after cutting, such as tops or branches, are discarded. Fig. 1 shows the material flow in rubber tree plantations in Cambodia. The volume of these residues is equal to 30% of the cut volume. The harvesting cost of the residues is lower than that of the logging residues in natural forests. Hence, the selling price is US $5–7/m$^3$, including the transportation cost to the customer. The residues are used as fuel for kilns at neighboring brick factories, which are strongly dependent on logging residues for fuel.

Clearcutting of pine (*Pinus merkusii*) is carried out every 35 years in the national forest in Central Java, Indonesia. The trees are cut by chainsaw, and the logs are skidded by human power. The tops and branches that are generated, as residues are less than 10 cm in diameter and account for 10% of the tree volume. The local community can acquire the residues free of charge and use them as residential fuelwood. The investigation was conducted in Java, which is heavily populated, and all logging residues appeared to be used in this manner, i.e., residues are collected manually.

At a large-scale, natural wood production site in Saba, Malaysia, 20–50 $m^3$/ha of cracked and hollow logs are discarded in the forest and in the landing[7]. In Kalimantan, Indonesia, 63.3 $m^3$/ha of residues are discarded in secondary forests after clearcutting[8]. Average logging residues in Asia represent 25–200% of log production[13], i.e., the residues represent 20–67% of the growing stock, and the residues from natural forest cutting appear to be equal to the log production volume[13]. However, the difficulty in using these residues is that the transportation cost of industrial residues is higher than their selling price. The former is approximately US \$20/$m^3$ [13] or US \$16–20/$m^3$ [7], whereas the latter is only US \$3–5/$m^3$ [7] in Malaysia. Thus, even if transportation were free of charge, the price disadvantage of the logging residues is considerable.

### 4.2 Industrial wood residues

We investigated the industrial wood residues in Cambodia[18, 24], Indonesia[17, 26, 27], Laos[28], and Malaysia.[13] The residues from PRT plantations are a major source of woody biomass in Cambodia[18] because the processing of forest trees has been limited by a logging ban. The tapping cycle of PRT is longer than in other countries, i.e., approximately 50 years. Thus, there are many thick logs from rubber trees. Fig. 1 also shows the material flow of rubber trees in the wood industry. The production yield of sawing is 43%, and the residues are mainly used as drying fuel for timber or are sold as fuel for brick kilns. The residues are partially used as packing materials, and 55% of the logs are used for energy production. All wood used as fuel for kilns in brick factories is derived from rubber trees.

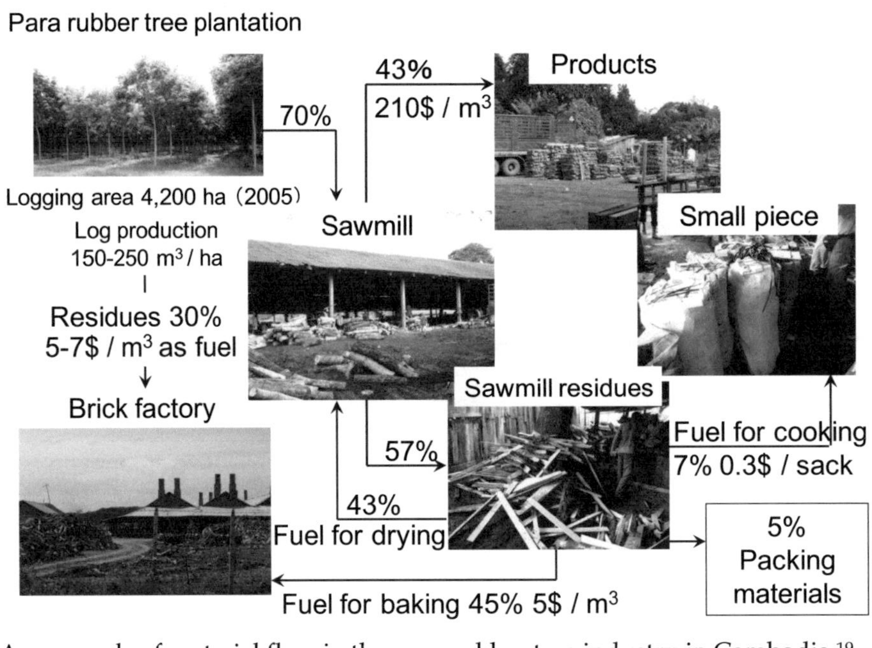


Fig. 1. An example of material flow in the para rubber tree industry in Cambodia [19]

Indonesia currently has 1,600 sawmills and 120 plywood factories[26]. For sawmills in Java, the production yield of teak (*Tectona grandis*) sawing is 50–60%, and the residues are used as fuel for drying ovens and by local people. For the plywood industry in the same area, the production yields from natural and planted wood are approximately 55% and 38%, respectively, because the average diameter of the former is approximately 70 cm, whereas that of the latter is less than 30 cm. One factory that we investigated used both natural and planted wood as raw materials and generated residues representing approximately 50% of raw wood consumption. A portion of these residues was used as raw material for laminated wood or blockboard. The rest of the residue was used as fuel to produce heat to dry veneer and for hot pressing. Some factories use both residues that are generated within and obtained from outside the factory to fuel combined heat and power supply (CHP) systems. Recently, the source of this raw material has been shifting from natural wood to planted wood because of the increase in natural wood prices and rising concerns about deforestation[17]. In Kalimantan, plywood factories are experiencing a shortage of large natural wood (Fig. 2). Most factories around Samarinda have ceased operations because of the decreasing amount of raw wood, which fell sharply from 150,000 $m^3$/month to 30,000 $m^3$/month[27]. Fig. 3 shows material flow in plywood factory A, not including materials other than wood such as adhesives.[29] The plywood yield was 50%, and this generated residue (primary residue). The primary residues consisted of sawdust, sander dust, log core, bark, log end and edges. The log cores were used for blackboard materials. Some log cores and edges were laminated and used for packing material. The remaining primary residues became secondary residues, and was used for fuel for the boiler. However, the factory lacked fuel that could be obtained there, and so it purchased wood fuel and brown coal externally. Fig. 4 and Table 4 show the energy consumption to produce plywood for the

Fig. 2. The raw wood in the plywood factory [19]

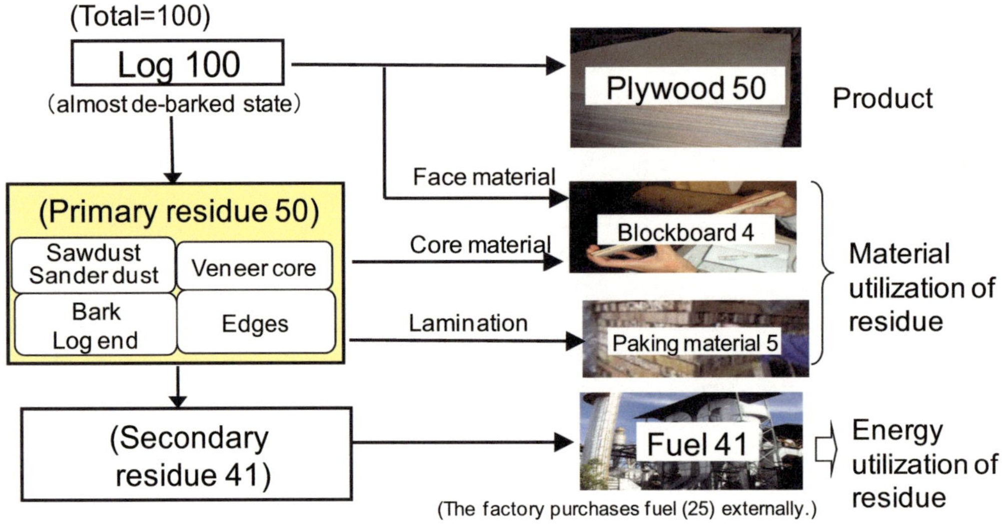


Fig. 3. Material flow in a plywood factory A in Indonesia [29]

(a)

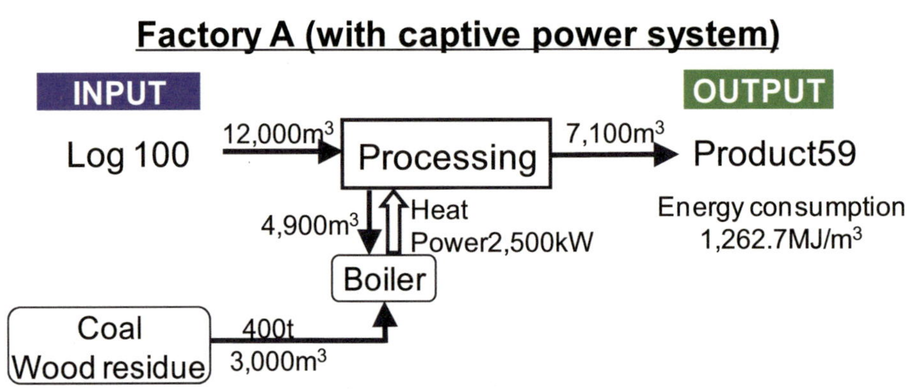


(b)

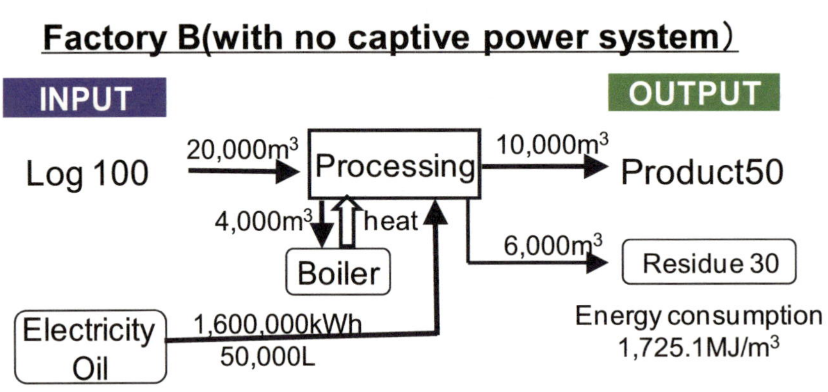


Fig. 4. Material and energy flow in plywood factory A (a) and B (b) in Indonesia [29]

| Factory | Fuel consumption (kL-gas oil) | Power consumption (kWh) | Total energy consumption ($MJ/m^3$-product) |
|---|---|---|---|
| A | 238 | 1,200,000 | 1,262.7 |
| B | 50 | 1,600,000 | 1,725.1 |

Table 4. Energy consumption for the plywood factory A and B in Indonesia
29: Reference source

plywood factories A and B. [29] The energy consumption was 1,263 and 1,725 $MJ/m^3$-product respectively. The smaller value for factory A was due to implementation of captive generation using wood fuel. According to reports, energy consumption during the making of plywood was 1,400 $MJ/m^3$-product. There is little significant difference between this data and our calculations.

Laos has 200 wood-processing factories, many of which are located near the capital Vientiane or in the southern part of the country. It has only one plywood factory. Charcoal factories are found everywhere because of the high demand of charcoal for residential use. Processing factories, which saw up to 15,000 $m^3$/year of logs for furniture, packing material, flooring, and doors, depend on wood from natural forests, including rosewood (*Dalbergia*) or meranti (*Shorea*), for raw materials. This natural wood is made available in the market in great quantities as a result of clearcutting for power dam development. The production yield in this industry is approximately 60%. All residues are used as fuel for boilers, and the sawdust is provided to the salt industry at no charge. However, in the southern part of Laos, unused sawing residues pile up in the backyards of sawmills.

In Malaysia, the total volume of the wood industry residues is 7.5 million $m^3$ annually[25]. Sawmill residues were highest in Sabah, and plywood mill residues were highest in Sarawak. The product yields in sawmills were 65% (western Peninsular Malaysia), 45% (Sabah), and 40% (Sarawak), and the yields in plywood mills were 50–60% (Peninsular Malaysia). The yields in molding factories were 74% (Sabah and Sarawak) and 70% (Peninsular Malaysia). According to a questionnaire conducted in the Kemena wood industry park of Sarawak by T.C. Wong (unpublished data), approximately 200,000 $m^3$ of logs were consumed per month and processed into plywood, timber, and fiberboard. The volume of wood residues was approximately 90,000 $m^3$ per month, of which 75% and 20% were generated by plywood mills and sawmills respectively. The percentage use of the original residues was 75%. Veneer core and slabs/offcuts were commonly used. In terms of application, steam/thermal use was dominant (48%), followed by fiberboard (40%), co-generation (7%), other use (4%), and manufacturing of charcoal briquettes (1%). Approximately 20,000 $m^3$ of residues were, however, disposed of directly or by incineration. Wong (unpublished data) noted that at least one additional briquette plant should be installed to use the remaining residues. In an industrial park in Bintulu, Sarawak, wood residues were sufficiently used as briquette charcoal, fiberboard material, and fuel for boilers. This was because of government incentives. To promote biomass use, the government remits duties in the implementation of related systems. In the Keningau area of Sabah, because wood-processing plants were small and separate from other plants, the residues, particularly sawdust, were not used and were usually incinerated at the plant. We estimated that the total unused sawdust remaining in the Sabah area is approximately 4,700 t/month.

## 5. Estimating the biomass energy potential

We estimated the biomass energy potential using the research data. We regarded all residues from the logging and wood industries as having biomass energy potential, although some residues had already been used as raw materials. The energy potential was estimated using the following equations[2]:

$$RV = MPV \times RPR/(100 - RPR) \quad (1)$$

$$EP = RV \times LHV \quad (2)$$

where RV is the residue volume ($m^3$), MPV is the main product volume ($m^3$), RPR is the residue production ratio (%), EP is the energy potential (toe), and LHV is the lower heating value (toe/$m^3$). Table 5 shows the EP of certain types of residues in some countries for which we were able to collect the required data. The main products of the PRT plantation residues and logging residues are logs. LHV is calculated from values stipulated by the International Energy Agency,[10] where the weight of 1 $m^3$ of wood is assumed to be 500 kg[10]. Cambodia has 0.16 Mtoe EP from PRT plantations and sawmills, accounting for approximately 6% of domestic energy consumption in 2001. In Thailand, PRT industries have the largest EP at 2.44 Mtoe (Table 5). Malaysia has an EP of 0.78 Mtoe from the logging residues in Sarawak, which is almost equal to the EP from sawmills.

It should be pointed out that ASEAN countries have high EP from oil palm industries[20, 21], which generate residues such as empty fruit bunches (EFB), fronds, and trunks, and a large portion of the EFB and almost all the frond and trunk residues are discarded in the plantation field. Table 6 shows the estimated amounts of EFB and fronds produced in Indonesia, Malaysia, and Thailand. Oil processing generates EFBs, and large quantities of fronds remain in the plantation area. A comparison to the data presented in Table 6 shows

| Item | | MPV ($Km^3$) | RPR (%) | RV ($Km^3$) | LHV ($GJ/m^3$) | EP (Mtoe) |
|---|---|---|---|---|---|---|
| Cambodia | PRT plantation residue | 840 [18, 24] | 30 [18, 24] | 360 | 7.4 [10] | 0.06 |
| | PRT sawmill | 361 [18, 24] | 57 [18, 24] | 479 | 8.4 [10] | 0.10 |
| Indonesia | Plywood | 3,353 [4] | 50 [17] | 3,353 | 8.4 [10] | 0.67 |
| | Sawmill | 4,330 [4] | 45 [17] | 3,543 | 8.4 [10] | 0.71 |
| Laos | Sawmill | 130 [4] | 40 [28] | 87 | 8.4 [10] | 0.02 |
| Malaysia | Plywood | 5,601 [4] | 50 [13] | 5,601 | 8.4 [10] | 1.12 |
| | Sawmill | 5,173 [4] | 45 [13] | 4,232 | 8.4 [10] | 0.85 |
| | Logging residue* | 4,440 [13] | 50 [13] | 4,440 | 7.4 [10] | 0.78 |
| Thailand | PRT sawmill | 10,800 [4] | 53 [21] | 12,179 | 8.4 [10] | 2.44 |

Table 5. Potential energy from woody biomass in selective ASEAN countries

MPV: Main product volume, RPR: Residue production ratio, RV: Residue volume, LHV: Lower heating value, EP: Energy potential

* The value is limited to Sarawak.

[4, 10, 13, 17, 18, 19, 21, 24, 28]: Reference source

| | | Moisture content wt% | Production Mton (dry) | Production Mton (wet) | LHV MJ/kg | EP Mtoe |
|---|---|---|---|---|---|---|
| EFB | Indonesia | 8.81 [3] | 3.46 [20] | 3.79 | 16.44 [3] | 1.49 |
| | Malaysia | | 4.71 [20] | 5.17 | | 2.03 |
| | Thailand | | 0.30 [20] | 0.33 | | 0.13 |
| Frond | Indonesia | 48.34 [3] | 24.14 [20] | 46.74 | 7.97 [3] | 8.90 |
| | Malaysia | | 32.93 [20] | 63.74 | | 12.14 |
| | Thailand | | 2.05 [20] | 3.97 | | 0.76 |

Table 6. Energy potential of palm oil industries in Indonesia, Malaysia, and Thailand
[3, 19, 20]: Reference source

that the oil palm industries have much larger EP than woody biomass. The problem remains that because the fronds cannot be used without transportation, preferable uses must be found either as a source of energy or as a source of material such as fiber. The residues from oil palm industries are certainly a valuable source of biomass with great potential.

## 6. Discussion

Here, we will discuss the difficulties of woody biomass use in the future. Although some districts produce massive quantities of logging residues, their use will hardly increase because of high transportation costs. The commercial collection and use of logging residues does not appear to have advanced for some time and is confined to regional uses such as cooking fuel for local residents. Hence, in regions that have small populations, a surplus of biomass is present compared to the quantity of available resources. The Philippines[23] and Vietnam[22] follow policies to promote the use of tree plantations, and Indonesia is promoting a shift in the source of raw materials to planted wood. Thus, the use of planted wood is increasing in the wood industries. Planted wood has a short-term harvesting cycle and produces small logs compared to natural forests. These small logs result in inferior yields of the main wood products. As a result, industrial residues will likely increase in the future, although it is expected that the shortage of wood materials will become serious because of the decrease in the number of large logs and because of advances in technology that uses the remaining residues for other purposes. Consequently, the residues used as raw materials for by-products will increase, whereas the residues that can be used as energy sources will decrease.

The use of fuelwood from industrial residues cannot be immediately replaced by an alternative. Because there are no equally low-cost alternatives for fuelwood from residues, these residues are necessary for use in drying ovens in the wood industry. The amount of residues was intentionally increased in some factories because of their need for sufficient fuelwood to supply the energy required for drying. Alternatives to the use of residues cannot be found unless the cost of residues rises and cheaper heat sources become available. Nevertheless, in some wood-processing factories, residues are discarded or incinerated. In such cases, the conversion of sawdust to briquette charcoal can be an efficient use of residues. In Laos and Cambodia, the implementation of CHP systems in wood-processing factories to generate electricity in local areas is promising. Government incentives are needed to develop such systems, as in the case of a wood-industry park in Sarawak, Malaysia.

Woody biomass is essential for local residents and for the wood industry and other industries in ASEAN countries. A possible approach to expand the uses for woody biomass as an energy source is as follows: First, the amount of biomass resources that are consumed should be reduced through the use of more efficient heat-using technology. Traditional cooking stoves are inefficient. If they are replaced with upgrades, the Philippines and Thailand could generate a surplus energy of 5.07 and 0.85 Mtoe (2005) respectively[2]. Woody biomass needs to be used as efficiently as possible, not only to save resources, but also to protect forests and precious ecosystems. Second, logging residues that are not collected because of current high transportation costs must be used. This problem cannot be resolved in the short term, but the whole world, including the ASEAN countries, must adopt this approach in the long run. ASEAN countries must deal with forestry problems such as illegal logging or improper slash-and-burn agriculture. A system needs to be put into place that will resolve these problems collectively. There is a need for a policy to encourage the collection of logging residues, especially those in the landings, as well as technology that will allow the efficient use of oil palm residues.

According to ACE estimates[1], biomass energy will play a significant role as a renewable energy source by 2010 in all ASEAN countries. It will account for 70% of renewable sources in 2010, although it accounted for 77% in 2000. Biomass energy will be a necessary energy source for the future in ASEAN countries, which must make concrete steps toward resolving the problems outlined in this study.

## 7. Acknowledgments

This work was partly supported by the Ministry of Education, Culture, Sports, Science and Technology of Japan (Feasibility Studies on Biomass in ASEAN Countries). We would also like to extend our thanks to Toshiro Harada, Masaki Jinkawa, Hirofumi Kuboyama, Motoe Miyamoto, Ryohei Tanaka, Eiji Togawa, Yoshiaki Tanaka, Takeshi Toma, Yuki Imatomi, Mario Tonosaki, Shozo Nakamura, Tadao Gotoh and Koichi Yamamoto for providing data and helpful comments.

## 8. References

[1] ACE (2002) Energy Statistics. *ASEAN Energy Bulletin*, 6(3), 19.

[2] Bhattacharya, S. C. et al. (2005) An assessment of the potential for non-plantation biomass resources in selected Asian countries for 2010. *Biomass & Bioenergy*, 29(3), 153-166.

[3] Boonrod, S. et al. (2005) Assessment of sustainable energy potential of non-plantation biomass resources in Thailand. *Biomass & Bioenergy*, 29(3), 214-224.

[4] FAO: FAOSTAT-forestry. Available online at http://faostat.fao.org/site/626/default.aspx#ancor.

[5] FAO. (2006) *Global Forest Resources Assessment 2005, Main Report. Progress Towards Sustainable Forest Management*. Food and Agriculture Organization of the United Nations, Rome, Italy, pp.320.

[6] Harada, T. et al. (2005) Stock and usage of wood biomass resources in Malaysia. *In* Proc. of the 14th Japan Institute of Energy Conference, Osaka, Japan, pp.152-153 [In Japanese].
[7] Harada, T. (2007) Topics of the 8th Pacific Rim Bio-Based Composites Symposium. *Mokuzai kogyo (Wood Industry)*, 62(6), 275-278 [In Japanese].
[8] Iskandar, H. et al. (2005) *Pemanfaatan Limbah Kayu Masyarakat. Report of ITTO Project PD 39/00 Rev. 3 (F)*, CIFOR, Indonesia, pp.29 [In Indonesian with English summary].
[9] IEA: IEA Statistics. Available online at http://www.iea.org/Textbase/stats/index.asp.
[10] IEA. (2007) *Good Practice Guidelines*. IEA Publications, France, pp.61.
[11] Joel, K. B. Jr. (2007) Biofuels: boon or boondoggle? *National Geographic*, 212(4), 38-59.
[12] Kouphokham, K. (2003) Energy planning and outlook of Lao PDR. *ASEAN Energy Bulletin*, 7(3), 6-8.
[13] Kuboyama, H., Tanaka, R. & Jinkawa, M. (2005) The utilization of woody biomass in Saba, Malaysia. *Mokuzai kogyo (Wood Industry)*, 60(5), 224-228 [In Japanese].
[14] Lean, T. (2003) Energy and electricity outlook of Cambodia. *ASEAN Energy Bulletin*, 7(3), 4-6.
[15] Malaykham, B. (2006) Brief country report of biomass in Lao PDR. *In* 3rd Biomass-Asia Workshop, Tsukuba, Japan, pp.39-46.
[16] National Statistics Center of the Lao PDR: Selected statistics. Available online at http://www.nsc.gov.la/Statistics/Selected%20Statistics/Population.htm.
[17] Suzuki, H. et al. (2006) Present situation of the woody biomass utilization in Java, Indonesia. *Nettai ringyo (Tropical Forestry)*, 67, 44-50 [In Japanese].
[18] Suzuki, H. et al. (2006) The utilization of woody residues in Cambodia. *In* 3rd Biomass-Asia Workshop, Tsukuba, Japan, pp.104.
[19] Suzuki, H. et al. (2009) Current Situation and Concerns with Woody Biomass Use in ASEAN Countries. Japan Agricultural Research Quarterly, 43 (1), 37-43.
[20] Tanaka, R. (2006) Woody biomass in south-east Asia-accumulation and availability. *Cellulose Commun*, 13(2), 61-65 [In Japanese with English summary].
[21] Tanaka, R. et al. (2006) Current situation of wood industry and woody biomass in Thailand. *Mokuzai kogyo (Wood Industry)*, 61(7), 286-292 [In Japanese].
[22] Yamamoto, K. et al. (2006) Availability of forest plantation resources as a biomass use in Vietnam. *In* Proc. of the 1st Biomass Science Conference, Tokyo, Japan, pp.62-63 [In Japanese with English summary].
[23] Yamamoto, K., Jinkawa, M. & Eusebio, D. A. (2007) Woody biomass and its utilization in the Philippines. *In* Proc. of the 2nd Biomass Science Conference, Hiroshima, Japan, pp.58-59 [In Japanese with English summary].
[24] Yamamoto, K. et al. (2007) Woody biomass utilization in ASEAN countries. *Journal of the Japan Institute of Energy*, 86, 375-379 [In Japanese with English summary].
[25] Yoshida, T. et al. (2005) Feasibility studies on woody biomass utilization in Malaysia and Indonesia. *In* International Conference on Coal Science and Technology, Okinawa, Japan, CD-ROM.
[26] Yoshida, T. et al. (2006) Availability of wood biomass resources in Indonesia. *In* Proc. of the 1st Biomass Science Conference, Tokyo, Japan, pp.64-65 [In Japanese with English summary].

[27] Yoshida, T. & Suzuki, H. (2006) The state of woody biomass in Indonesia. *Mokuzai kogyo (Wood Industry)*, 61(9), 404-407 [In Japanese].
[28] Yoshida, T. et al. (2007) Availability of woody biomass resources in Laos. *In* Proc. of the 2nd Biomass Science Conference, Hiroshima, Japan, pp.56-57 [In Japanese with English summary].
[29] Yoshida, T. et al. (2009) Energy Flow Analysis on Logging and Processing of Plantation Wood in Indonesia. Japanese Journal of International Forest and Forestry, No.75, 38-44 [in Japanese].

# 6

# Electricity Generation by Photosynthetic Biomass

Chun-Chong Fu, and Wen-Teng Wu
*Department of Chemical Engineering, National Cheng Kung University, Tainan 70101, Taiwan*

## 1. Introduction

Biofuel cells generate electricity through biological processes. Conventional microbial fuel cells (MFCs) operate by converting organic substrates, such as glucose, acetate, starch, and lactate, to electrical bioenergy through microbial oxidation processes [1-3]. A typical MFC consists of an anodic and a cathodic chamber with electrodes partitioned by a proton exchange membrane (PEM) or a cation exchange membrane. This membrane functions as an insulator for maintaining the redox potential and only allows specific ion exchange [4].
While MFCs use suspension cultivation of microorganisms in the anodic chamber [5], some MFCs attach microorganisms to the electrodes to form a biofilm [6]. As microbial oxidation consumes the supplied substrates, the anode surface generates electrons and conducts them to the cathode through an external circuit. The resulting cations pass through the membrane to the cathode in the electrolyte. However, the following three factors limit MFC performance: (1) electron activation on the anode and cathode surfaces, (2) electron transfer from microbial cells to the anode, and (3) internal resistances of the circuit and anions passing through the membrane. Researchers have developed electrode modification, mediator addition, and membrane-free designs to address these issues and improve MFC performance [5]. Different electrode materials produce different activation polarization losses; for example, the noble metal platinum (Pt) offers superior catalytic activity. But, graphite, graphite felt, Pt-coated graphite, and other metal-coated materials are employed as cost-effective electrodes [5,7,8].
Since the cell surfaces of microorganisms are not electrically conductive, the electrons inside the cells cannot directly transfer to the surrounding electrolyte [9,10]. For this reason, previous designs adopt several kinds of toxic and unstable electro-chemicals as electrochemical mediators, or electron shuttles: neutral red (NR), methylene blue (MB), thionint, and phenolic compounds [4,10]. However, these active chemicals are expensive and toxic, rendering them unsuitable for long-term operation [4]. Exchange membranes are the most expensive components of typical MFC [10]. These membranes insulate and separate different kinds and/or concentrations of electrolytes into two chambers and restrict specific ion exchange. Once the permeability of the exchange membrane becomes poor, the resulting increase in ohmic resistance decreases MFC performance [11]. A membrane-free system may address these concerns. In a membrane-free system, the diffusion gradient of the dissolved reactants between the anode and cathode maintains the electrochemical

potential with a sufficiently long distance [5]. However, this sufficient long distance leads to a large internal resistance [4,10].

Many operational MFCs employ microbes such as *Clostridium* sp., *Geobacter* sp., *Shewanella* sp., *Synechocystis* sp. and some photosynthetic microorganisms as microbial catalysts for electricity generation [4, 12, 13]. These non-photosynthetic and photosynthetic microorganisms require external substrates to live and generate electricity in the heterotrophic cultivation mode. Several recent studies apply mediator-free and membrane-free MFCs for energy recovery in wastewater treatment [10, 11, 14, 15]. This promising design for MFC is also applied in our study. The current study uses the photosynthetic microorganism *Spirulina platensis* for electricity generation. When *S. platensis* attaches to the anode, it maintains the chemical potential using the attached biomass and conducts the produced electrons to the anode directly instead of transporting them through a mediator, exchange membrane, or reactant gradient. Both photosynthetic reaction in light and respiration reaction in dark can generate electrical power. The attractive design factors of a photosynthetic microbial fuel cell (PMFC), not only include its mediator-free and membrane-free design, but also instant electricity generation instead of a long cultivation time [16]. This study attempts to operate a PMFC in buffer solution using the batch mode with no organic substrate. This study also examines the effects of electrode spacing, electrolyte pH, temperature, and light on OCV associated on the PMFC power output.

## 2. Materials and methods

### 2.1 Cultivation of *Spirulina platensis*

*S. platensis* was obtained from the Fisheries Research Institute in Pintung, Taiwan, and it was cultivated in Zarrouk medium containing 16.00 g/L $NaHCO_3$, 1.60 g/L NaCl, 1.00 g/L KCl, 0.50 g/L $K_2HPO_4$, 0.2 g/L $MgSO_4.7H_2O$, 0.10 g/L $(NH_2)_2CO$, 0.04 g/L $CaCl_2{\cdot}2H_2O$, and 0.01 g/L $FeSO_4{\cdot}7H_2O$ [17]. Cultivation was conducted in a 2000 ml Duran bottle with a 1700 ml working volume. The aeration rate of air was 2LPM and a white fluorescent lamp provided a continuous illumination of 30 μmole photon $m^{-2}$ $s^{-1}$. The cultivation system was operated at temperatures ranging from 28 to 30 °C for 7 days before *S. platensis* was harvested and weighed for the experiments.

### 2.2 Configuration of a membrane-free and mediator-free PMFC

The proposed PMFC chamber was constructed of transparent poly-acrylic plastic material. The length, width, and height of the PMFC chamber were 4cm, 2cm, and 2cm, respectively. Two platinum electrodes measuring 5cm long, 0.5cm wide, and 0.1cm thick were installed at each end of the PMFC with 2cm and 4cm spacing. One of the electrodes served as the anode for *S. platensis* attachment. Medium solution, carbonic acid, sodium bicarbonate, and sodium carbonate were employed as electrolytes. Using the medium as the electrolyte prepared the mediator-free and membrane-free PMFC for use, as Fig. 1 shows.

## 3. PMFC operation and measurements

The experiments in this study operated the PMFC in a continuous light illuminated and temperature controlled environment. *S. platensis* with biomass weight of 0.2g was attached to the 0.25$cm^2$ anode. The PMFC's voltage and current readings were collected every 10 seconds by a dual-channel multi-meter (PROVA 903, manufactured in Taiwan) connected to the Pt electrodes and a laptop PC through a RS-232 interface. Experiments used different temperatures (20°C, 30°C, and 40°C), light conditions (0 and 30μmol $m^{-2}$ $s^{-1}$), electrolyte pH values (pH 5.5, 8.5 and 9.9), and spacing between electrodes (2cm and 4cm) to determine their effects on the output voltage and current. Resistances of 22MΩ, 10MΩ, 3.9MΩ, 1MΩ, 500KΩ, 220KΩ, 100KΩ, 56KΩ, 18KΩ, 1KΩ, and 2.5Ω were loaded to reveal the relationship between working current and PMFC voltage in the external circuit. Furthermore, two equally-sized PMFCs were connected in series and in parallel to investigate the effect of setup on overall performance.

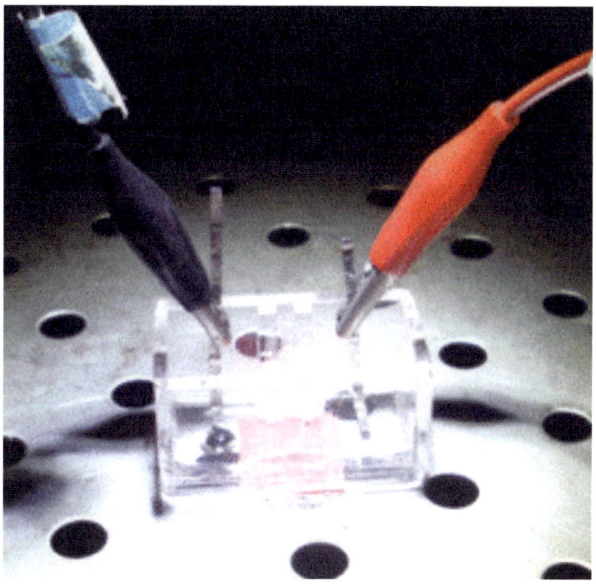

Fig. 1. The mediator-free and membrane-free PMFC with *S. platensis*, which attached to the left side.

## 4. Results and discussion

### 4.1 Initialization and instant application

Initially, 0.2g *S. platensis* biomass weight was attached to the anode surface under dark conditions, and the distance between the electrodes was 4cm. Since the biomass attaching onto the anode is a step change for the PMFC, the OCV increased from zero to 0.33V in 13 minutes after the biomass attachment. Figure 2(a) shows the associated responses to the mentioned step change in the OCV readings. Equation 1 indicates that the OCV transition time depends on the time constant (*RC*), which is the product of the internal resistance (*R*) and capacitance (*C*) of the PMFCs:

$$RC = \frac{-t}{\ln(1-\frac{V}{V_m})} \tag{1}$$

where *R* is the internal resistance of the PMFC, *C* is the capacitance, $V$ is the voltage, *t* is the time, and $V_m$ is the maximum OCV.

Applying the least squares method to the recorded OCV data indicates that the time constant, *RC*, is approximately 4.4 minutes. The time constant represents the time required to reach 63.2% maximum OCV. Reducing the internal resistance decreases the time constant decreases, and simultaneously increases power output [18]. The time constant indicates that the developed PMFC is capable of instant usage.

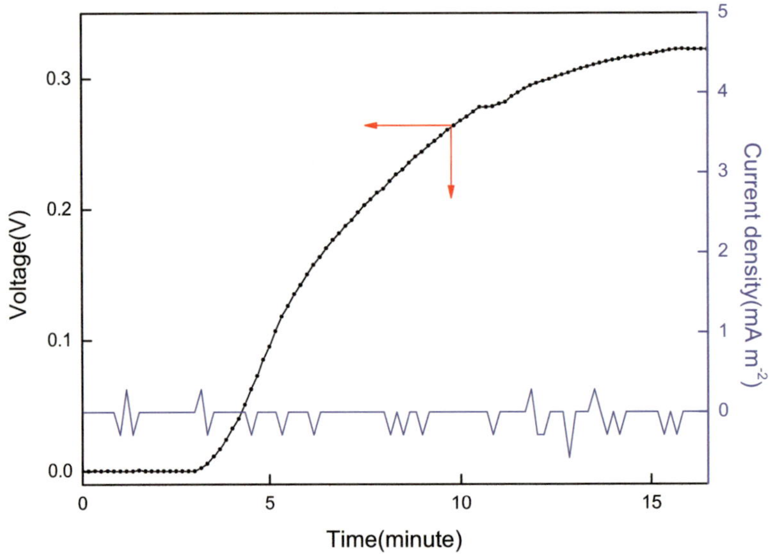


Fig. 2.(a) The response of the OCV associated with the zero current density caused by a step change of 0.2g biomass attachment on the anode.

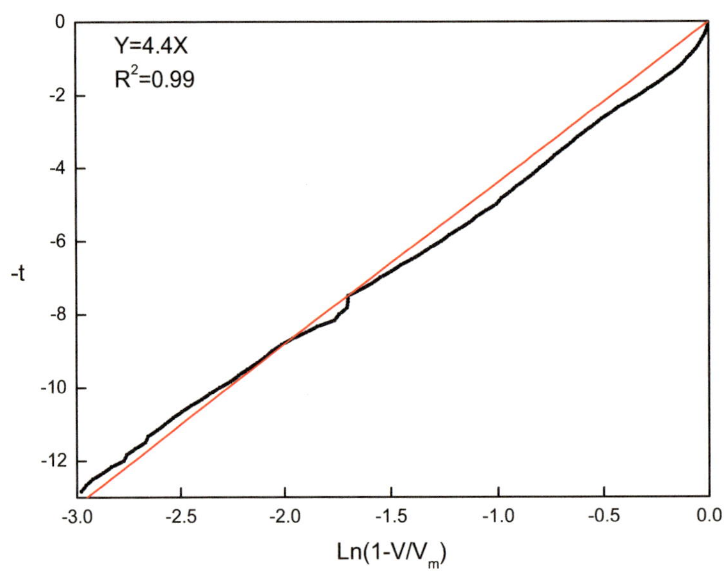


Fig. 2.(b) Determination of the time constant, *RC*, by the least squares method with the collected OCV data.

### 4.2 Effects of light conditions

The proposed *S. platensis* PMFC was operated in the dark until the OCV approached a pseudo-steady-state level in the first eight minutes. A light intensity of 30μmol photon $m^{-2}$ $s^{-1}$ was applied to the PMFC. The PMFC responses the step change coming from the light intensity by the OCV dropped from 0.24V to 0.19V, as shown in Fig. 3. This negative response in lighting is different from other studies, indicating that the OCV increased with the light intensity [13,19].

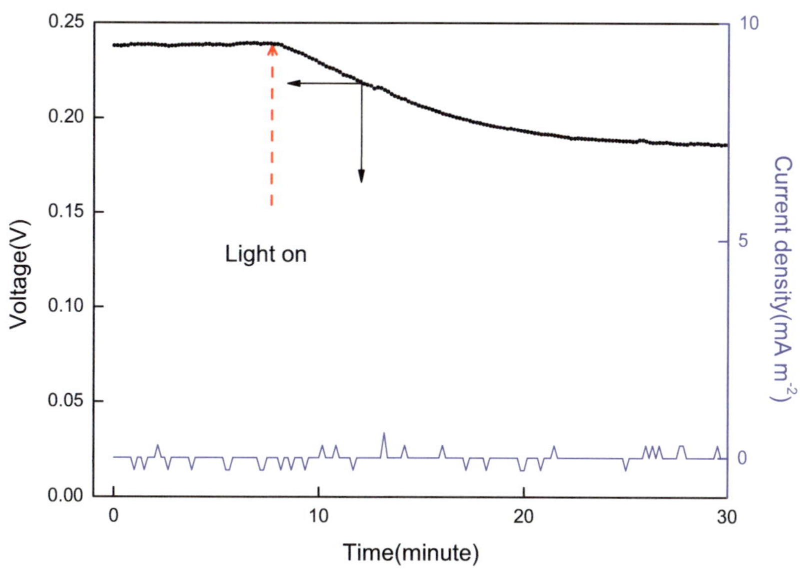


Fig. 3. Effects of lighting on the OCV and output current density.

### 4.3 Effects of resistance

The PMFC was connected to various levels of external resistances to generate electricity in both dark and light conditions. The OCV of the PMFC was initially 0.24V in the dark condition. A step change of 22MΩ resistor created a voltage drop of 0.22V, and the current density increased from zero to 0.3mA $m^{-2}$. Resistances of 10MΩ, 3.9MΩ, 1MΩ, 500KΩ, 220KΩ, 100KΩ, 56KΩ, 18KΩ, 1KΩ, and 2.5Ω, were sequentially applied to determine the resistance effects on PMFC voltage and current densities. The resistance change took about five minutes to approach a new pseudo-steady-state level. Decrease of the external resistanceed led to a decrease of working voltages and an increase of current densities until voltage readings approaching zero, as shown in Fig. 4.

### 4.4 Effects of spacing in dark and light conditions

Spaces between the electrodes were provided to evaluate the output of the voltages and current densities under both the light and dark conditions. The PMFC with 4cm electrode spacing was first operated in dark and light conditions, and responded to the applied light with an OCV drop. Various levels of external resistances were sequentially loaded to determine the associated current densities and voltages in dark conditions. Results indicate that a shorter spacing distance created higher voltage, current density, resistance, and output. The PMFC under light conditions achieved a lower power density in the same external resistance than that under dark condition.

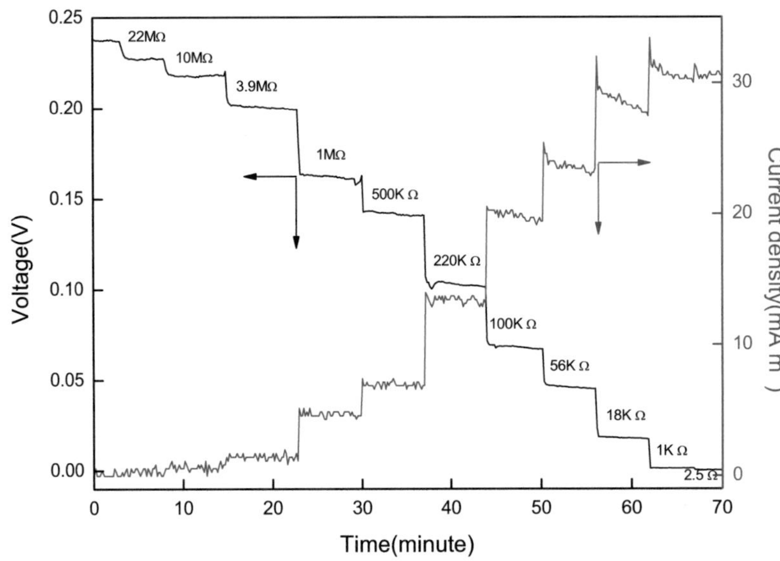


Fig. 4. Time courses of voltage associated with the current density readings of a PMFC with 4 cm electrode spacing after sequentially loading various resistances.

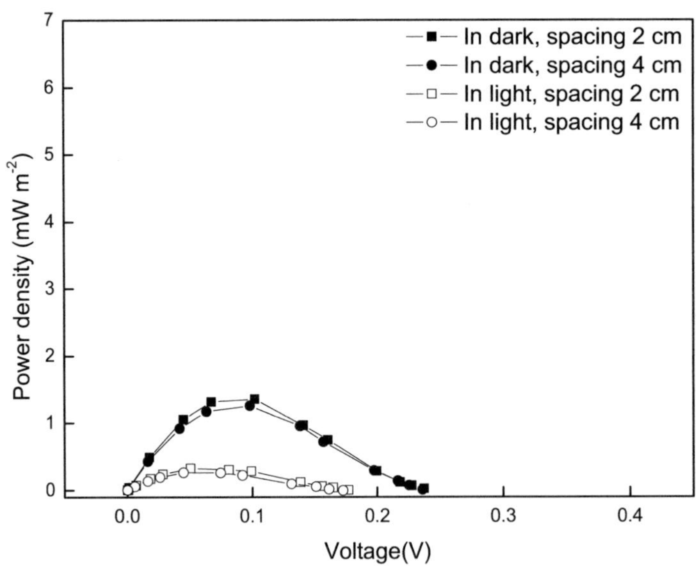


Fig. 5. Power density curves of the PMFC with different electrode spaces of 2 cm and 4 cm under dark and light conditions (pH 9.9, 30°C, and biomass density of 1g $cm^{-2}$).

### 4.5 Effects of electrolyte pH

Three pH levels of 5.5, 8.3, and 9.9, maintained by carbonic acid, sodium bicarbonate and sodium carbonate, respectively, were applied to the PMFC to evaluate the effects of pH on power output. Experimental results demonstrate that the highest OCV of 0.39V occurred at pH 5.5; while the lowest OCV of 0.24V occurred under basic conditions (pH 8.3 and pH 9.9) The received maximum power output were approximately 5mWm$^{-2}$, as Fig 6 shows. These results indicate that the PMFC performed better in acidic conditions. A possible explanation is that the carbonic acid increased the PMFC's ionic strength of $H^+$ and reduced the internal resistance of PMFC [20].

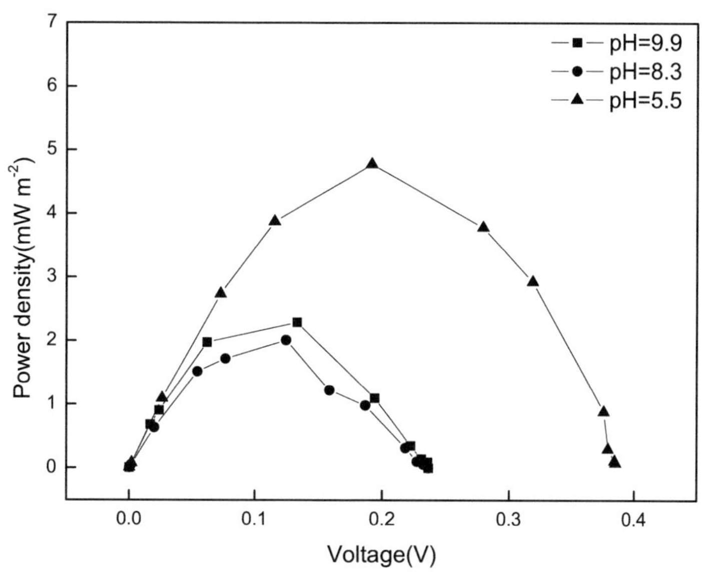


Fig. 6. Power density curves of the PMFC with *S. platensis* at various pH levels of 5.5, 8.3, and 9.9 (in dark, 30°C, and biomass density of 1g cm$^{-2}$).

### 4.6 Effects of temperature

The PMFC was operated at 20°C, 30°C, and 40°C to determine effects of temperature on electrical output. The OCV increased as the temperature increased, and the maximal value of 0.39V appeared at 40°C. Figure 7 shows the power density curves of the PMFC loaded with various external resistors. These results show that PMFC achieved higher power output at higher temperatures. A possible explanation is that higher temperatures increased the reaction rate and transportation of electrons [21-23].

### 4.7 Effects of PMFC connections

Since the negative light response in this study differs from other studies, subsequent experiments examined the effects on the connection of two PMFCs in parallel and in series. External resistors were loaded sequentially to obtain voltage and current density readings. The resulting current-voltage curves in Fig. 8 present that the OCV readings for parallel and series connections were 0.31V and 0.45V, respectively, and the maximum current densities were 40 and 25mA m$^{-2}$. These results indicate that PMFCs connected in parallel and series achieved greater current densities and OCVs, respectively. The maximal power density was approximately 2.5mW $m^{-2}$ for both cases.

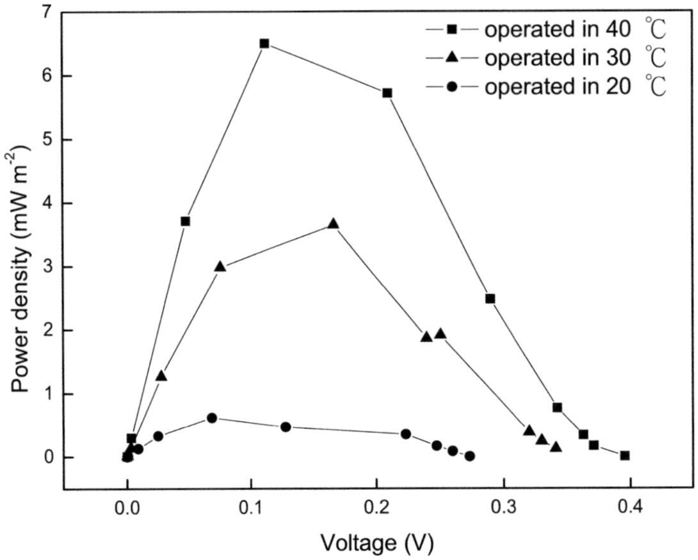


Fig. 7. Power density curves of the PMFC with *S. platensis* under temperature conditions of 20°C, 30°C, and 40°C (in dark, pH 9.9, and biomass density of 1g $cm^{-2}$).

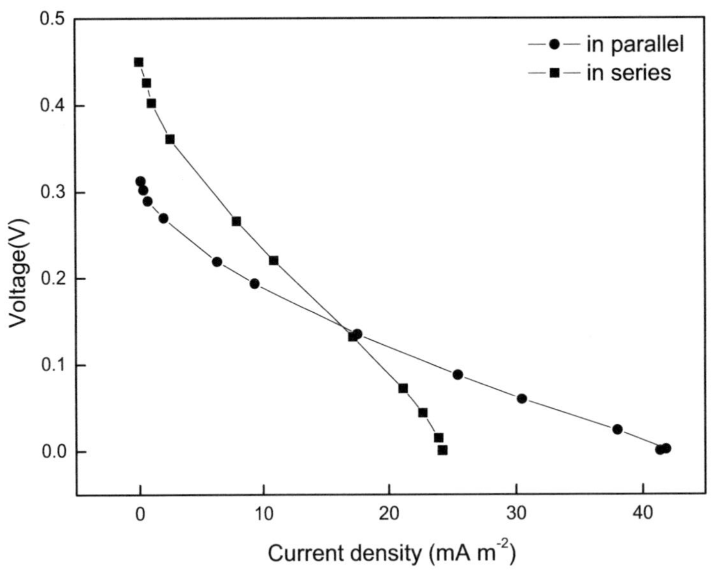


Fig. 8. The polarization curves of two equally-sized PMFCs connected in series and in parallel (in dark, pH 9.9, 30°C, and biomass density of 1g $cm^{-2}$).

## 5. Conclusion

The proposed PMFC employs the living bio-catalyst *S. platensis* to generate electricity without membranes and mediators. This study examines PMFC performance under different lighting conditions, electrode spaces, electrolyte pH values, temperatures, and connection types. The proposed PMFC achieved the highest power output in the conditions of dark, 2cm between the electrodes, pH 5.5, and a temperature of 40°C. When two PMFCs of the same size were connected, they exhibited a higher voltage in series and greater current density in parallel.

## 6. Reference

[1] Z.W. Du, H.R. Li, T.Y. Gu, A state of the art review on microbial fuel cells: A promising technology for wastewater treatment and bioenergy, Biotechnology Advances, 25 (2007) 464-482.

[2] B.E. Logan, B. Hamelers, R. Rozendal, U. Schröder, J. Keller, S. Freguia, P. Aelterman, W. Verstraete, K. Rabaey, Microbial Fuel Cells: Methodology and Technology, Environmental Science & Technology, 40 (2006) 5181-5192.

[3] K. Rabaey, W. Verstraete, Microbial fuel cells: novel biotechnology for energy generation, Trends in Biotechnology, 23 (2005) 291-298.

[4] P. Clauwaert, K. Rabaey, P. Aelterman, L. De Schamphelaire, T.H. Ham, P. Boeckx, N. Boon, W. Verstraete, Biological denitrification in microbial fuel cells, Environmental Science & Technology, 41 (2007) 3354-3360.

[5] Z. Du, H. Li, T. Gu, A state of art review on microbial fuel cells: A promising technology for wastewater treatment and bioenergy, Biotechnology Advances, 25 (2007) 464-482.

[6] D.R. Bond, D.R. Lovely, Electricity production by Geobacter sulfurreducens attached to electrodes, Apply Enviroment Microbiology, 69 (2003) 1548-1555.

[7] C. Dumas, A. Mollica, D. Feron, R. Basseguy, L. Etcheverry, A. Bergel, Checking graphite and stainless anodes with an experimental model of marine microbial fuel cell, Bioresource Technology, 99 (2008) 8887-8894.

[8] S.Q. Yang, B.Y. Jia, H. Liu, Effects of the Pt loading side and cathode-biofilm on the performance of a membrane-less and single-chamber microbial fuel cell, Bioresource Technology, 100 (2009) 1197-1202.

[9] F. Davis, S.P.J. Higson, Biofuel cells-Recent advances and applications, Biosensors and Bioelectronics, 22 (2007) 1224-1235.

[10] J.K. Jang, T.H. Pham, I.S. Chang, K.H. Kang, H. Moon, K.S. Cho, B.H. Kim, Construction and operation of a novel mediator- and membrane-less microbial fuel cell, Process Biochemistry, 39 (2004) 1007-1012.

[11] M.M. Ghangrekar, V.B. Shinde, Performance of membrane-less microbial fuel cell treating wastewater and effect of electrode distance and area on electricity production, Bioresource Technology, 98 (2007) 2879-2885.

[12] D.R. Bond, D.E. Holmes, L.M. Tender, D.R. Lovely, Electrode-reducing microorganisms that harvest energy from marine sediments, Science, 295 (2002) 483-485.

[13] K. Tanaka, R. Tamamushi, T. Ogawa, Bioelectrochemical fuel-cells operated by the cyanobacterium, *Anabaena variabilis*, Journal of Chemical Technology and Biotechnology 35B (1985) 191-197.

[14] L.P. Huang, R.J. Zeng, I. Angelidaki, Electricity production from xylose using a mediator-less microbial fuel cell, Bioresource Technology, 99 (2008) 4178-4184.
[15] B. Min, B.E. Logan, Continuous electricity generation from domestic wastewater and organic substrates in a flat plate microbial fuel cell, Environmental Science & Technology, 38 (2004) 5809-5814.
[16] C.-C. Fu, C.-H. Su, T.-C. Hung, C.-H. Hsieh, D. Suryani, W.-T. Wu, Effects of biomass weight and light intensity on the performance of photosynthetic microbial fuel cells with Spirulina platensis, Bioresource Technology, 100 (2009) 4183-4186.
[17] C.-Y. Wang, C.-C. Fu, Y.-C. Liu, Effects of using light-emitting diodes on the cultivation of *Spirulina platensis*, Biochemical Engineering Journal, 37 (2007) 21-25.
[18] T.L. Floyd, Electronics Fundamentals- Circuit, Devices, and Applications, Pearson Education, Upper Saddle River, NJ, 2004.
[19] Y.J. Zou, J. Pisciotta, R.B. Billmyre, I.V. Baskakov, Photosynthetic Microbial Fuel Cells With Positive Light Response, Biotechnol. Bioeng., 104 (2009) 939-946.
[20] H. Liu, S. Cheng, L. Huang, B.E. Logan, Scale-up of membrane-free single-chamber microbial fuel cells, Journal of Power Sources, 179 (2008) 274-279.
[21] G.S. Jadhav, M.M. Ghangrekar, Performance of microbial fuel cell subjected to variation in pH, temperature, external load and substrate concentration, Bioresource Technology, 100 (2009) 717-723.
[22] S.W. Hong, I.S. Chang, Y.S. Choi, T.H. Chung, Experimental evaluation of influential factors for electricity harvesting from sediment using microbial fuel cell, Bioresource Technology, 100 (2009) 3029-3035.
[23] B. Min, O.B. Roman, I. Angelidaki, Importance of temperature and anodic medium composition on microbial fuel cell (MFC) performance, Biotechnology Letter, 30 (2008) 1213-1218.

7

# Microalgae-based Systems for Carbon Dioxide Sequestration and Industrial Biorefineries

Eduardo Jacob-Lopes[1] and Telma Teixeira Franco[2]
[1]*School of Agricultural Engineering, Federal University of Pelotas, UFPel, 96010-900, Pelotas-RS,*
[2]*School of Chemical Engineering, State University of Campinas, UNICAMP, P.O. Box 6066, 13083-970, Campinas-SP, Brazil*

## 1. Introduction

The bulk of the evidence indicating that global climatic alterations occur as a result of increasing concentrations of greenhouse gases in the atmosphere has created pressure to develop strategies to reduce these changes (IPCC, 2001). Carbon dioxide is considered to be the main gas of the greenhouse effect, both in terms of emission and its climate-altering potential.

In 1997, the signatory countries of the Kyoto Protocol agreed to reduce $CO_2$ emissions in an agreement that established the need to develop carbon dioxide sequestering processes. Thus the various technologies available for carbon capture and storage need to be evaluated from the point of view of obtaining carbon credits, aiming to stabilize emissions of this pollutant (UNFCCC, 1997). In addition to technologies available for immediate use, other $CO_2$ capture methods are being developed for application in the near future. The choice of these methodologies will depend on factors such as cost, capture capacity, environmental impact and the speed with which the technology can be introduced in addition to social factors such as public acceptance (IPCC, 2007a).

In this context, the use of biotechnological processes for carbon dioxide biofixation is considered viable for reducing emissions of this pollutant. These processes are based on the use of reactors used to develop photosynthetic reactions in which microalgae are used as biocatalysts in a series of biochemical reactions responsible for the conversion of $CO_2$ into photosynthetic metabolic products (Jacob-Lopes et al., 2010). With this in mind, the objectives of this present chapter are to present an overview of a potential technology for carbon dioxide transformation into biomolecules and to describe the current state of the art in the biological conversion of $CO_2$ in photobioreactors thereby facilitating worldwide advances in this research area.

## 2. Carbon dioxide emissions

Global monitoring of atmospheric $CO_2$ concentration during the last century indicated an increase in carbon dioxide concentration from 295ppm in 1900 to 377ppm in 2004, representing an increase of 27.8% (Thitakamol et al., 2007). On a global basis, it is estimated that more than 25 $GtCO_2$ are emitted annually as a result of burning fossil fuels. The magnitude of the influence of human activities on the biological carbon cycles suggests the need for high managerial levels and the mitigation of emissions of this compound into the atmosphere (IPCC, 2007b).

Sources of carbon dioxide emission can be classified as stationary, mobile or natural. The industrial processes most contributing to increasing atmospheric $CO_2$ concentrations consist of electrical energy generating plants, hydrogen and ammonia production plants, cement factories, and fermentative and chemical oxidation processes. In addition to the carbon dioxide emitted industrially, the $CO_2$ generated in residences, buildings and commercial complexes also contributes to the stationary emissions, as do forest and agricultural fires. The mobile emission sources mainly consist of the carbon dioxide generated by passenger and cargo transport including cars, trucks, buses, planes, trains and ships. Human and animal metabolism, plant and animal degradation and volcanic and oceanic activities are the main natural carbon dioxide sources. Sources of anthropogenic emissions include stationary and mobile sources but exclude the natural sources (Song, 2006).

Microalgae-based systems are restricted to the use of stationary industrial emissions. Sources of high purity $CO_2$ emission at reduced temperatures should be identified and the photobioreactors adapted to these conditions (Francisco et al., 2010).

## 3. Microalgae

Current taxonomic concepts and standards classify microalgae into groups as diatoms, chlorophyceae and cyanobacteria (Anand, 1998).

Photosynthesis is the main metabolic model of the microalgae, a process that had a central role in the rise in the oxygen level of the terrestrial atmosphere during the evolution of the current biosphere (Schmetterer, 1994). Nevertheless these microorganisms have great versatility in the maintenance of their structures, using different energy metabolisms such as respiration and nitrogen fixation (Demeyer et al., 1982; Grossman et al., 1994).

Some genera of microalgae have high concentrations of pigments, including chlorophyll *a*, considered essential for photosynthesis. Another two pigment classes involved in light energy capture are the carotenoids and phycobilins. The carotenoids are red, orange or yellow lipid-soluble pigments, found in association with chlorophyll *a*. The third class of accessory pigments is the phycobilins: phycocyanin, a blue pigment present in microalgae, and phycoerythrin, a red pigment sometimes absent (Fay, 1983). In addition to these pigments, these microorganisms have a highly developed intracytoplasmatic system, indicating photosynthesis as the preferred metabolic pathway.

The microalgae are capable of using free $CO_2$ and bicarbonate ions as a source of inorganic carbon during photosynthesis, transporting them across the fine plasmatic membrane where they accumulate in the cell as an inorganic carbon reservoir for photosynthesis. The bicarbonate is converted into $CO_2$ by the enzyme carbonic anhydrase (Zak et al., 2001; Badger & Price, 2003).

The main characteristic of photosynthesis, first elucidated in algae and higher plants, can also be applied to the microalgae, although there are some aspects specific to some microalgae. The spectral light absorption characteristic of these strains is different from that of the other photosynthetic organisms, since high photosynthetic activity rates are measured not only in the spectral region from 665 to 680nm, where the light is better absorbed by chlorophyll *a*, but also from about 620nm to 560nm, where phycocyanin and phycoerythrin respectively absorb light effectively. This shows that the light absorbed by the phycobiliproteins is used by these microalgae as efficiently as light absorbed by chlorophyll, suggesting a very high photosynthetic activity by these microorganisms (Campbell et al., 1998).

### 3.1 Photosynthetic metabolism

Photosynthesis is characterized by a two-stage mechanism: a photochemical reaction and a carbon fixation reaction. In this way, carbon dioxide is incorporated into ribulose 1,5 diphosphate (rubisco) energy being required during the catalytic reaction of the primary enzyme rubisco carboxylase. The reaction product is broken into three carbon molecules, phosphoglyceric acid (PGA) and the reduction of the PGA caused by the electron transporter NADPH (nicotinamide adenine dinucleotide phosphate) leads to the production of a series of intermediary phosphorylated sugars and finally to glucose. This sequence of metabolic transformations is known as the Calvin-Benson cycle (Calvin and Benson, 1948).
Carbon dioxide fixation is not directly light dependent and thus the process is called the photosynthetic dark reaction. The demands for energy in the form of ATP and NADPH translate the transformations of the Calvin-Benson cycle, entirely dependent on the photochemical reaction, which occurs in the tilacoid or intracytoplasmatic membrane (Campbell et al., 1998). In this stage the light energy is absorbed by the highly organized structures of the photosynthetic pigments and electron transporters, known as photosystems I and II, thus exciting the chlorophyll *a* molecule. This leads to an explosion of excited electrons and their flow determines the redox potential gradient, which results in the formation of strongly electronegative electron transporters such as ferridoxin and NADPH. Part of the energy liberated is incorporated into ATP in the phosphorylation process during electron transport. The last electron source for photosynthesis is $H_2O$, which gives up hydrogen atoms and electrons during the photolysis process, or Hill's reaction, and releases $O_2$, the product of photosynthesis by microalgae and green plants (Fromme et al., 2006).
Although carboxylation by rubisco is the main $CO_2$ incorporation pathway in microalgae under optimum photosynthesizing conditions, this is not the only carbon dioxide fixation pathway. The carboxylation of phosphenol pyruvate, catalyzed by the enzyme phosphenol pyruvate carboxylase, is another $CO_2$ fixation pathway. Oxaloacetate is easily converted into $C_4$ dicarboxylic acids, for example into malate or citrate, and subsequently into amino acids such as aspartate or glutamate. This pathway, left over from the $C_4$ dicarboxylic acid pathway in higher plants, complements the pentose phosphate-reducing pathway in microalgae. The presence of two carboxylation systems, operating in parallel, could represent an important adaptation of the microalgae to sharp environmental changes. Under limited light conditions, carbon assimilation is preferentially channeled in the direction of the synthesis of amino acids and other essential cell constituents, but under saturated light conditions, sugars and starch are formed via the pentose phosphate-reducing pathway. This indicates that with intense illumination, the $CO_2$ fixation rate can exceed the rate of nitrogen assimilation and, thus, the excess carbon and energy derived from photosynthesis are stored in the form of glycogen (Fay, 1983; Campbell et al., 1998; Zak et al., 2001).
The dark endogenous metabolism serves mainly as an agent for the photosynthetic and biosynthetic mechanisms for the subsequent active light period. Glycogen is the main reserve product, which can support limited dark metabolism and provide the energy

maintenance required for essential cell processes in the dark. It is first converted into glucose-6-phosphate, which is then metabolized via the respiratory pathways (Fay, 1983). Although enzymes from the glycolytic pathway have been identified in microalgae, they show extremely low activity. The energy metabolism of the dark metabolism of the microalgae is distinctly dependent on $O_2$ and its main pathway is the pentose-6-phosphate oxidative cycle (Schmetterer, 1994).

### 3.2 Carbon concentration mechanisms in microalgae

The way in which the different species of microalgae adapt to a wide range of carbon dioxide concentrations is related to an essential biophysical mechanism denominated the carbon concentration mechanism (CCM), which concentrates the carbon dioxide at the photosynthetic carboxylation sites. This mechanism corresponds to complex metabolic pathways, since different forms of inorganic carbon are involved in these biological processes (Jaiswal & Kashyap, 2002).

The function of the carbon concentration mechanism is to raise the intracellular inorganic carbon levels, compensating for limitations in the carbon dioxide supply that could reduce the photosynthetic rates. This mechanism is responsible for pumping $CO_2$ to the carboxylation sites (Falkowski, 1997).

Microalgae are capable of using three different inorganic carbon assimilation pathways: (i) direct carbon dioxide assimilation via the plasmatic membrane; (ii) the use of bicarbonate by inducing the enzyme carbonic anhydrase, which converts the $HCO_3^-$ into $CO_2$; and (iii) direct transport of bicarbonate via the plasmatic membrane. The enzymes carbonic anhydrase and ribulose 1,5 bisphosphate carboxylase/dehydrogenase (rubisco) are responsible for the biocatalysis of these reactions, in which the enzyme carbonic anhydrase converts bicarbonate into carbon dioxide, and rubisco uses this compound as a substrate to produce phosphoglycerate. The rate of this reaction may be slow due to limited carbon dioxide production. Thus the elevated efficiency of the enzyme carbonic anhydrase, capable of increasing the intracellular carbon dioxide levels to concentrations 1000 times higher than those in the external fluid, results in an efficiency carbon fixation reaction in these organisms. These mechanisms are consistent with various results found in the literature about microalgae with high carbon dioxide requirements and capable of accumulating high internal levels of inorganic carbon (Fridlyand et al., 1996; Marcus, 1997; Tchernov et al., 1997; Badger & Price, 2003; Cuaresma et al., 2006).

## 4. Photobioreactors

Biotechnological processes have been conducted in the evaluation of the mass and energy transference phenomena, in the dimensioning and construction of equipment processing biotransformations and in the operation of control systems and instrument applications for accompanying the transformation kinetics (Merchuk & Wu, 2004).

Photobioreactors using microalgae to treat polluting compounds and produce biomolecules are based on five basic criteria: elevated efficiency in the use of light energy, an adequate mixing system, easy control of the reaction conditions, reduced hydrodynamic stress on the cells and ease in scale-up (Muñoz & Guieysse, 2006).

Systems using photobioreactors are based on natural processes in which the photosynthetic metabolism of the microorganisms converts light energy, heat and $CO_2$ into photosynthetic products (Contreras et al., 1999). The use of photobioreactors to cultivate microalgae requires the presence of light, carbon dioxide and dissolved nutrients for growth of the microorganism. Consequently, these processes require systems for illumination, gas exchange (addition of $CO_2$ and removal of $O_2$), the addition of nutrients and temperature control (Rorrer & Cheney, 2004).

Photosynthetic microorganisms can be cultivated in open or closed photobioreactors. The closed systems are characterized by elevated photosynthetic efficiency associated with a precise control of the operational variables, showing a lower risk of contamination and minimization of water loss by evaporation, highly significant factors in open systems. On the other hand, closed systems are more expensive, since they must be constructed with transparent materials, and are more complicated to operate and more difficult to scale up. The ratio of volume per unit area is another criterion to be considered when choosing the system, since the implementation of open systems requires the availability of large areas for the elevated reaction volumes (Borowitzka, 1999; Molina Grima et al., 1999).

Various configurations have been proposed for closed photobioreactors. The main types include bubble column, air-lift, tubular (loop) and stirred tank reactors (Jacob-Lopes et al., 2009). Open pond systems can be oval (raceway), circular or rectangular (Borowitzka, 1999).

### 4.1 Carbon dioxide transfer in photobioreactors

Carbon dioxide is usually the main carbon source in the photosynthetic cultivation of microalgae and can be transferred continually or intermittently from the gas phase to the liquid phase of the culture medium (Molina Grima et al., 1999).

The reactivity of carbon dioxide in aqueous solutions establishes various equilibriums in its contact with water. The first equilibrium refers to the dissolution of the gas in the water, forming carbonic acid. The carbonic acid undergoes almost instantaneous dissociation into bicarbonate and carbonate ions with the total inorganic carbon concentration being given by the sum of the species $CO_3^{2-}$, $HCO_3^-$ and $CO_2$ (Rorrer & Mullikin, 1999).

Simple $CO_2$ bubbling in the liquid phase does not lead to a total dissolution, since a fraction of the injected $CO_2$ is lost in the gas outlet. $CO_2$ absorption is mainly a function of the volumetric mass transfer coefficient, the mass transfer driving force and the gas retention time (Merchuk et al., 2000).

In terms of solubility, carbon dioxide is approximately ten times more soluble in water than oxygen gas. Nevertheless, due to the low solubility of both gases in aqueous solution, there is a need to provide these elements throughout the process (Klasson et al., 1991).

Thus an efficient carbon dioxide transfer system is required for photobioreactors. Efficiency in carbon dioxide transfer is necessary so as to raise the volumetric mass transfer coefficients $K_La$ ($CO_2$) allowing for improved transfer of gas to the liquid phase (Baquerisse et al., 1999). According to these authors, the volumetric mass transfer coefficients depend mainly on the physical properties of the fluid, the fluid flow and the system and geometry of the gas injector.

Carbon dioxide transfer in bioreactors becomes a limiting factor in the processes, since the dissolved carbon dioxide concentration decreases with increase in temperature and also with an increase in the concentration of dissolved salts. This factor is relevant in processes for the transfer and removal of $CO_2$ by microalgae, suggesting the need for higher values of saturation concentration (Rorrer & Cheney, 2004).

## 5. Carbon dioxide biotransformation by microalgae

Microalgae are microorganisms that are being applied in the reduction of carbon dioxide emissions into the atmosphere, where this compound is biotransformed in the presence of light energy. Evidence of a highly developed photosynthetic system has led to the suggestion of using microalgae in the treatment of gaseous effluents with elevated $CO_2$ concentrations, generated by industrial discharges (Hsueh et al., 2007).

Much research was developed in the nineties, especially in Japan, on processes for the biofixation of carbon dioxide using microalgae. These studies are being intensified, aiming at projecting systems that operate efficiently and economically with the objective of developing technologies for the reduction of gaseous pollutants (Watanabe & Hall, 1996; Watanabe & Saiki, 1997; Cheng et al., 2006; Ono & Cuello, 2007; Jacob-Lopes et al., 2008; Jacob-Lopes et al., 2009, Francisco et al., 2010).

The use of microalgae in carbon dioxide convertion processes is considered a promising alternative, since the element carbon can be converted by different mechanisms. In the first step, the carbon dioxide dissolved in the aqueous phase of the system can be sequestered by chemical precipitation due to the reaction of the ions bicarbonate and carbonate with elements present in the culture medium, such as calcium and magnesium. These reactions are catalyzed by the growth and physiology of the microalgae (Marcus, 1997; Lee et al., 2004). Another carbon-fixing pathway is related to the Calvin-Benson cycle, where specialized enzymes present in these organisms catalyze reactions that incorporate carbon atoms coming from the $CO_2$ involved in photosynthesis (Falkowski, 1997). The biological conversion of carbon dioxide results in products of the photosynthetic metabolism such as cells, oxygen, biopolymers soluble in the culture medium and volatile organic compounds (VOC's) (Ishida et al., 1997; Muñoz et al., 2004; Jacob-Lopes et al., 2010).

The $CO_2$ conversion into biomass is high only under conditions where the $CO_2$ mass loading rate is low. At a high $CO_2$ mass loading rate, the formation of volatile organic compounds is the main $CO_2$ biotransformation route (Fig. 1).

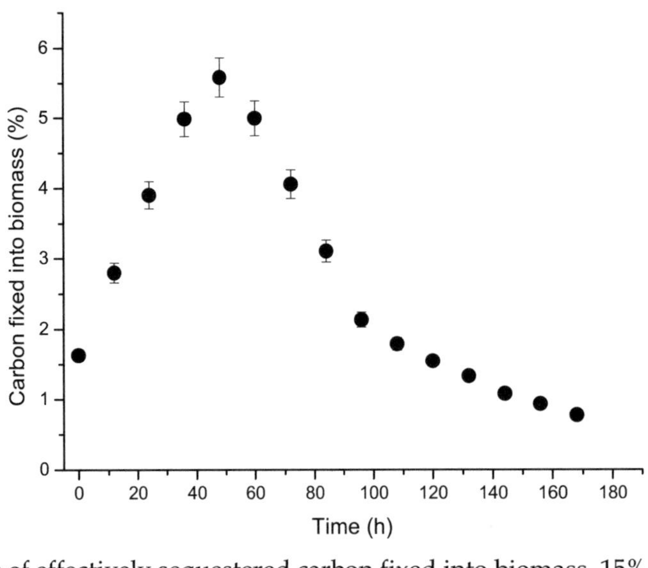


Fig. 1. Percentage of effectively sequestered carbon fixed into biomass. 15% of $CO_2$ at a flow rate of 1VVM. Source: Jacob-Lopes et al. (2010).

## 6. Potential uses for the bioproducts

The main advantages of producing biomolecules from photosynthetic organisms are related to their rapid reproduction and the low cost of the sources of energy and nutrients used for their multiplication. Evidently this is done for the cost of the medium in which the microorganisms develop, which can be composed of a wide variety of substrates, some of which, including industrial residues, are cheap, thus solving problems of an environmental nature and also serving to produce consumables (Anupama & Ravindra, 2000).

The biochemical composition of the microalgal cells includes characteristics of commercial interest, significant proportions of proteins, lipids, carbohydrates and pigments, which can be used as ingredients of foods destined for human consumption and animal feeds and in the extraction of biomolecules and the production of biofuels (Harun et al., 2010). In this way, the use of these microorganisms in carbon sequester processes associates the treatment of polluting compounds with the production of consumables that can be recycled in a variety of forms. Table 1 shows some potential uses for the bioproducts formed by the biological conversion of carbon dioxide in photobioreactors.

An analysis of Table 1 demonstrates a wide variety of possible uses for the microalgal biomass. According to Spolaore et al. (2006), the microalgal biomass industry currently produces more than 5000 tons of dried mass/year with an annual revenue greater than US$ $1.25x10^9$, not including processed products, demonstrating the exploration potential of this type of biotechnological process.

| Application | Examples | Reference |
|---|---|---|
| Human food | Source of single-cell protein and use in the supplementation of products such as pastas, soups and beverages. | Rodriguez-Garcia and Guil-Guerrero (2008) |
| Animal feed | Frequent use of some species in the feeding of fish and shellfish. | Olvera-Novoa et al. (1999) |
| High-value molecules | Source of chlorophyll *a*, phycocyanin, β-carotene, γ-linolenic acid, eicosapentaenoic acid and stable isotope biochemicals | Spolaore et al. (2006) |
| Fertilizers | Use of the biomass as a source of nitrogen and phosphorous in tillable land. | Chae et al. (2006) |
| Natural gas production | Production of $CH_4$ in fermenters by the digestion of biomass. | Yen and Brune (2007) |
| Biodiesel production | Production of biodiesel from the lipid fraction of the cells. | Miao and Wu (2006) |
| Syngas production | Production of synthesis gas from the biomass. | Amin (2009) |
| Inorganic salts production | Source of carbonates and bicarbonates | Lee et al. (2004) |
| Renewable production polymers | Source of exocellular sugars and proteins. | Ishida et al. (1997) |
| Volatile organic compounds production | Production of hydrocarbons, aldehydes and organohalogens | Muñoz et al. (2004) |

Table 1. Potential uses of the bioproducts

Besides the use of biomass and its derivatives, carbonates and bicarbonates are other products likely to be formed in photobioreactors. The use of Generally Recognized as Safe (GRAS) species and airstreams without toxic compounds, e.g., bioethanol plants, can produce chemicals of commercial value (Huijgen et al., 2007).

In addition, extracellular proteins and mainly sugars can be secreted into culture media in photobioreactors. Such compounds have several applications in pharmaceutical and food industries, since some may have unique properties for special applications, not found in the polymers currently available. These include use as a bioemulsifier, bioflocculant, agar-agar substitute or cosmetic material as well as other (De Philippis & Vicenzini, 1998).

Finally, microalgae cells can produce non methane hydrocarbon (ethane, ethylene, propane, propylene, butane, isobutane, pentane, hexane, isoprene and ethylene) (Schobert & Elstner, 1980; Shaw et al., 2003), organohalogens (chloroform, trichloroethylene, bromomethane, chloromethane, iodomethane) (Scarratt & Moore, 1996) and aldehydes (propanal, hexanal, n-heptanal, formaldehyde, acetaldehyde, furfural and valeraldehyde) (Schobert & Elstner, 1980; Nuccio et al., 1995). These compounds are continuously being formed and released from the aqueous phase of photobioreactors. The production of renewable polymers is an emergent area for industrial practice.

Thus, microalgae-based systems are one the most promising emerging biorefinery platforms. These systems are a means of resolving environmental problems and providing effective solutions to the energy crisis at the same time. This biorefinery type mediates between environment and society and has positive economic impacts.

## 7. Applicability of the process

Full-scale processes with microalgae are mainly based on open photobioreactors. Some successful initiatives have been carried out in closed systems, and in this case, the scale is normally semi-pilot or pilot. The intensification of these processes represents an important step in the consolidation of the technology for the biological transformation of carbon dioxide into photosynthetic products.

Open photobioreactors are suitable for the production of high-value products. The carbon dioxide sequestration rates are low and are not viable for processes that aim only to obtain carbon credits.

Closed photobioreactors have higher rates of biotransformation of carbon dioxide into bioproducts, and the greatest potential for commercial application. However, there are still high hurdles to overcome before these processes can be fully scalable.

Although it is believed that there is high availability of industrial $CO_2$, for use in microalgae-based systems in the practice this is not true. Biologically mediated processes require cold gases, rarely obtained in the conventional industrial flue gases, which can reach thousands of degrees Celsius, limiting the use of most sources of industrial gases.

In addition, flue gases are usually composed of carbon dioxide and water vapor as well as nitrogen and excess oxygen remaining from the intake combustion air. They also contains a significant percentage of other compounds such as particulate matter, carbon monoxide, nitrogen oxides and sulfur oxides, which have a toxic and/or inhibitor effect on microalgal growth.

Finally, there is no consensus on the shape of closed photobioreactors for full-scale application. Bubble column, air-lift, flat-panel and tubular reactors and variations of these are the main options, but they are still far from industrial reality.

## 8. References


Amin, S., 2009. Review on biofuel oil and gas production processes from microalgae. Energy Conversion and Management 50, 1334-1349.

Anand, N., 1998. Cyanobacterial Taxonomic - Classical Concepts and Modern Trends. In: G. Subramanian, B.D. Kaushik and G.S. Venkataraman, Editors, Cyanobacterial Biotechnology, Science Publishers Inc., USA, pp. 337–340.

Anupama, P., Ravindra, L., 2000. Value-added food: Single cell protein. Biotechnol. Adv. 18, 459-479.

Badger, M.R., Price, G.D., 2003. $CO_2$ concentration mechanisms in cyanobacteria: Molecular components, their diversity and evolution. J. Exp. Bot. 54, 383, 609-622.

Baquerisse, D., Nouals, S., Isambert, A., Santos, P.F., Durand, G., 1999. Modeling of a continuous pilot photobioreactor for microalgae production. J. Biotechnol. 70, 335-342.

Borowitzka, M.A., 1999. Commercial production of microalgae: Ponds, tanks, tubes and fermenters. J. Biotechnol. 70, 313-321.

Calvin, M., Benson, A.A., 1948. The path of carbon in photosynthesis. Science. 107, 476-480.

Campbell, D., Hurry, V., Clarke, A.K., Gustafsson, P., Öquist, G., 1998. Chlorophyll fluorescence analysis of cyanobacterial photosynthesis and acclimation. Microbiol. Mol. Biol. R. 30, 667-680.

Chae, S.R., Hwang, E.J., Shin, H.S., 2006. Single cell protein production of *Euglena gracilis* and carbon dioxide fixation in an innovative photo-bioreactor. Bioresource Technol. 97, 322-329.

Cheng, L., Zhang, L., Chen, H., Gao, C., 2006. Carbon dioxide removal from air by microalgae cultured in a membrane-photobioreactor. Sep. Purif. Technol. 50, 324-329.

Contreras, A., Garcia, F., Molina Grima, E., Merchuk, J.C., 1999. Influence of sparger on energy dissipation, shear rate, and mass transfer to sea water in a concentric-tube airlift bioreactor. Enzyme Microb. Tech. 25, 820-830.

Cuaresma, M., Garbayo, I., Vega, J.M., Vílchez, C., 2006. Growth and photosynthetic utilization of inorganic carbon of the microalga *Chlamydomonas acidophila* isolated from Tinto river. Enzyme Microb. Tech. 40, 158-162.

De Philippis, R., Vincenzini, M., 1998. Exocellular polysaccharides from cyanobacteria and their possible applications. FEMS Microbiology Reviews 22, 151-175.

Demeyer, A., Jacob, F., Menguy, G., Perrier, J., 1982. La Conversion Bioenergetique-Durayonnement Solaire et le Biotechnologies. Chap. 2, Paris: Ed. Lavoisier. p. 276-301.

Falkowski, P.G., 1997. Photosynthesis: The paradox of carbon dioxide efflux. Curr. Biol. 7, 637-639.

Fay, P., 1983. The Blue Greens (Cyanophyta - Cyanobacteria), 5$^{th}$ Ed. London, Ed. Edward Arnold, pp. 88.

Francisco, E.C., Neves, D.B, Jacob-Lopes, E., Franco, T.T., 2010 Microalgae as feedstock for biodiesel production: Carbon dioxide sequestration, lipid production and biofuel quality. Journal of Chemical Technology and Biotechnology 85, 395-403.

Fridlyand, L., Kaplan, A., Reinhold, L., 1996. Quantitative evaluation of the role of a putative $CO_2$-scavenging entity in the cyanobacterial $CO_2$-concentrating mechanism. BioSystems. 37, 229-238.

Fromme, P.Y.H., Deruyter, Y.S., Jolley, C., Chauhan, D.K., Melkozernov, A., Grotjohann, I., 2006. Structure or photosystems I and II. CR Chim. 9, 188-200.

Grossman, A.R., Schaefer, M.R., Chiang, G.G., Collier, J.L., 1994. The Responses of Cyanobacteria to Environmental Conditions: Light and Nutrients. In: Bryant, D.A. The Molecular Biology of Cyanobacteria. Kluwer Academic Pubishers. pp. 641-668.

Harun, R., Singh, M., Forde, G.M., Danquat, M.K., 2010. Bioprocess engineering of microalgae to produce a variety of consumer products. Renewable and Sustainable Energy Reviews 14, 1037-1047.

Huijgen, W.J.J, Comans, R.N.J., Witkamp, G.J., 2007. Cost evaluation of $CO_2$ sequestration by aqueous mineral carbonation, Energy Conversion and Management 48, 1923-1935.

Hsueh, H.T., Chu, H., Yu, S.T., 2007. A batch study on the bio-fixation of carbon dioxide in the absorbed solution from a chemical wet scrubber by hot springs and marine algae. Chemosphere 66, 878-886.

IPCC, Intergovernmental Panel on Climate Change, 2001. The scientific basis. http://www.ipcc.ch/.

IPCC, Intergovernmental Panel on Climate Change, 2007a. Carbon dioxide capture and storage. http://www.ipcc.ch/.

IPCC, Intergovernmental Panel on Climate Change., 2007b. Mitigation of climate change. http://www.ipcc.ch/.

Ishida, T., Hasegawa, N., Hayashi, N.R., Peerapornpisal, Y., Ishii, M., Igarashi, Y., Kodama, T., 1997. Growth characteristics and dense culture of a thermophilic cyanobacterium *Chroococciciopsis* sp. strain TS-821. J. Ferment. Bioeng. 83, 6, 571-576.

Jacob-Lopes, E., Revah, S., Hernández, S., Shirai, K., Franco, T.T., 2009. Development of operational strategies to remove carbon dioxide in photobioreactors. Chemical Engineering Journal 153, 120-126.

Jacob-Lopes, E., Lacerda, L.M.C.F., Franco, T.T., 2008. Biomass production and carbon dioxide fixation by *Aphanothece microscopica Nägeli* in a bubble column photobioreactor. Biochem. Eng. J. 40, 27-34.

Jacob-Lopes, E., Scoparo, C.H.G., Queiroz, M.I., Franco, T.T., 2010. Biotransformations of carbon dioxide in photobioreactors. Energy Conversion and Management 51, 894-900.

Jaiswal, P., Kashyap, A., 2002. Isolation and characterization of mutants of two diazotrophic cyanobacteria tolerant to high concentrations of inorganic carbon. Microbiol. Res. 157, 83-91.

Klasson, K.T., Ackerson, M.D., Clausen, E.C., Gaddy, J.L., 1991. Bioreactor design for synthesis gas fermentations. Fuel, 70, 605-614.

Lee, B.D., Apel, W.A., Walton, M.R., 2004. Screening of cyanobacterial species for calcification. Biotechnol. Progr. 20, 1345-1351.

Marcus, Y., 1997. Distribution of inorganic carbon among its component species in cyanobacteria: Do cyanobacteria in fact actively accumulate inorganic carbon? J. Theor. Biol. 185, 31-45.

Merchuk, J.C., Wu, X., 2004. Simulation of algae growth in a bench scale internal loop airlift reactor. Chem. Eng. Sci. 59, 2899-2912.

Merchuk, J.C., Gluz, M., Mukmenev, I., 2000. Comparison of photobioreactors for cultivation of the red microalga *Porphyridium* sp. Journal of Chemical Technology and Biotechnology 75, 1119-1126.

Miao, X., Wu, Q., 2006. Biodiesel production from heterotrophic microalgal oil. Bioresource Technol. 97, 841-846.

Molina Grima, E., Fernandéz, F.G.A., Camacho, F.G., Chisti, Y., 1999. Photobioreactors: Light regime, mass transfer, and scale up. J. Biotechnol. 70, 231-247.

Muñoz, J., Mudge, S.M., Sandoval, A., 2004. Effects of ionic strength on the production of short chain volatile hydrocarbons by *Dunaliella salina* (Teodoresco). Chemosphere 54, 1267-1271.

Muñoz, R., Guieysse, B., 2006. Algal-bacterial processes for the treatment of hazardous contaminants: A review. Water Res. 40, 2799-2815.

Nuccio, J., Seaton, P.J., Kieber, R.J., 1995. Biological production of formaldehyde in the marine environmental. Limnol. Oceanogr. 40(3), 521-527.

Olvera-Novoa, M.A., Domýnguez, L.J., Olvira-Castillo, L., Martýnez-Palacios, C.A., 1999. Effect of the use of the microalgae *Spirulina maxima* as fish meal replacement in diets for tilapia, *Oreochromis mossambicus* P. fry. Aquacult. Res. 71, 219-225.

Ono, E., Cuello, J.L., 2007. Carbon dioxide mitigation using thermophilic cyanobacteria. Biosystems Eng. 96, 129-134.

Otero, A., Vincenzini, M., 2003. Extracellular polysaccharide synthesis by *Nostoc* strains as affected by N source and light intensity. J Biotechnol.102, 143-52.

Rodriguez-Garcia, I., Guil-Guerrero, J.L., 2008. Evaluation of the antioxidant activity of three microalgal species for use as dietary supplements and in the preservation of foods. Food Chem. 108, 1023-1026.

Rorrer, G., Cheney, D., 2004. Bioprocess engineering of cell and tissue cultures for marine seaweeds. Aquacult. Eng. 32, 11-41.

Rorrer, G.L., Mullikin, R.K., 1999. Modeling and simulation of a tubular recycle photobioreactor for macroalgal cell suspensions cultures. Chem. Eng. Sci. 54, 3153-3162.

Scarratt, M.G., Moore, R.M. 1996. Production of methyl chloride and methyl bromide in laboratory cultures of marine phytoplankton. Marine Chemistry 54, 263-272.

Schmetterer, G., 1994. Cyanobacterial Respiration. In: Bryant, D.A. The Molecular Biology of Cyanobacteria. Kluwer Academic Plubishers. pp.409-435.

Schobert, B., Elstner, E.F., 1980. Production of hexanal and ethane by *Phaeodactylum triconutum* and its correlation to fatty acid oxidation and bleaching of photosynthetic pigments. Plant Physiol. 66, 215-219.

Shaw, S.L., Chisholm, S.W., Prinn, R.G., 2003. Isoprene production by *Prochlorococcus*, a marine cyanobacterium, and other phytoplankton. Marine Chemistry 80, 227-245.

Song, C., 2006. Global challenges and strategies for control, conversion and utilization of $CO_2$ for sustainable development involving energy, catalysis, adsorption and chemical processing. Catal. Today 115, 2-32.

Spolaore, P., Cassan, C.J., Duran, E., Isambert, A., 2006. Commercial applications of microalgae. J. Biosci. Bioeng. 101, 87-96.

Tchernov, D., Hassidim, M., Luz, B., Sukenik, A., Reinhold, L., Kaplan, A., 1997. Sustained net $CO_2$ evolution during photosynthesis by marine microorganisms. Curr. Biol. 7, 723-728.

Thitakamol, B., Veawab, A., Aroonwilas, A., 2007. Environmental impacts of absorption-based $CO_2$ capture unit for post-combustion treatment of flue gas from coal-fired power plant. Int. J. Greenhouse Gas Control 1, 318-342.

UNFCCC. Kyoto Protocol to the United Nations Framework Convention on Climate Change, 1997. http://unfccc.int/resource/docs/convkp/kpeng.pdf.

Watanabe, Y., Hall, D., 1996. Photosynthetic $CO_2$ conversion technologies using a photobioreactor incorporating microalgae energy and material balances. Energ. Convers. Manage. 37, 1321-1326.

Watanabe, Y., Saiki, H., 1997. Development of a photobioreactor incorporating Chlorella sp. for removal of CO2 in stack gas. Energ. Convers. Manage. 38, 499-503.

Yen, H.W., Brune, D.E., 2007. Anaerobic co-digestion of algal sludge and waste paper to produce methane. Bioresource Technol. 98, 130-134.

Zak, E., Norling, B., Maintra, R., Huang, F., Andersson, B., Pakrasi, B., 2001. The initial steps of biogenesis of cyanobacterial photosystems occurs in plasma membranes. Plant Biology 32, 13443-13448.

# 8

# Production of Chemicals from Selective Fast Pyrolysis of Biomass

Xi-feng Zhu and Qiang Lu
*University of Science and Technology of China*
*P.R.China*

## 1. Introduction

With the increasing concern on fossil fuel storage and environmental problems, the utilization of renewable lignocellulosic biomass resources will play an increasing important role in the future. Biomass can be converted to a variety of fuels and chemicals by different technologies, one of them is fast pyrolysis which has received extensive interests in recent years (Bridgwater & Peacocke, 2000; Mohan et al., 2006). Fast pyrolysis of biomass is a thermal decomposition process that occurs in the absence of oxygen, with quickly biomass decomposition and rapid vapor condensation, to convert biomass mainly into a liquid product (know as bio-oil) with the yield as high as 70~80 wt%. The essential principles to obtain high bio-oil yield include moderate pyrolysis temperature (around 500 °C), very high heating rates ($10^3$ - $10^5$ °C/s), short vapor residence time (< 2 s) and rapid quenching of pyrolysis vapors. A number of pyrolysis reactors have been developed for the bio-oil production, including bubbling fluidized bed, entrained bed, circulating fluidized bed, rotating cone, screw pyrolysis reactor, vacuum pyrolysis reactor, etc. In recent years, several research institutes (BTG, Dynamotive) have already established demonstration biomass fast pyrolysis plants, suggesting that the fast pyrolysis technique is near commercial.

Crude bio-oils, also referred to as biomass pyrolysis liquids, pyrolysis oils, or bio-crude oils, are dark brown, free flowing liquids with an acrid or smoky odor. Chemically, bio-oils are complex mixtures of water and hundreds of organic compounds that belong to acids, aldehydes, ketones, alcohols, esters, anhydrosugars, furans, phenols, guaiacols, syringols, nitrogen compounds, as well as large molecular oligomers (holocellulose-derived anhydro-oligosaccharides and lignin-derived oligomers). There are many valuable compounds, and thus, bio-oils have the potential for useful chemicals recovery. However, most of the chemicals are in low contents, making their recovery not only technically difficult but also economically unattractive at present. Therefore, the commercialization of bio-oils for value-added chemicals requires the production of specific bio-oils with high contents of target products.

Selective fast pyrolysis, differed from conventional fast pyrolysis which is usually aimed at the maximum bio-oil yield, is to drive the pyrolysis of biomass towards the products of interest. Since the biomass pyrolytic pathways and the subsequent products are influenced by various factors, and thus, selective fast pyrolysis can be performed in different ways, mostly by catalyst utilization, to (1) maximize the yield of target products and (2) obtain target products with high purity. This manuscript is divided into two sections. The first section reviews the biomass fast pyrolysis reaction pathways and mechanisms, and the second section discusses the methods for the production of several value-added chemicals from selective fast pyrolysis of biomass.

## 2. Biomass fast pyrolysis reaction pathways and mechanisms

### 2.1 Cellulose fast pyrolysis

Cellulose is the main component of lignocellulosic biomass and is predominantly located in the cell wall. It is a linear homopolysaccharide of β-D-glucopyranose units linked together by (1→4)-glycosidic bonds. Among the three major components of lignocellulosic biomass, the cellulose has received the most attention in its pyrolytic mechanism study.

Cellulose starts pyrolysis at as low as 150 °C. At temperatures lower than 300 °C, pyrolysis of cellulose mainly involves the reduction in degree of polymerization, the formation of free radicals, elimination of water, formation of carbonyl, carboxyl and hydroperoxide groups, and evolution of carbon monoxide and carbon dioxide, finally leaving a charred residue (Shafizadeh, 1982; Evans & Milne, 1987). The low temperature pyrolysis will produce very low yield of organic liquid products.

At temperatures above 300°C, the pyrolysis of cellulose involves many new reactions, mainly leading to a liquid product with the yield as high as 87 wt% (Piskorz et al., 1989). Generally, cellulose is firstly decomposed to form activated cellulose (Boutin et al., 1998). Afterwards, two major parallel pathways will take place, the depolymerization and the fragmentation (ring scission), as shown in Fig.1. The depolymerization process mainly forms anhydro-oligosaccharides, levoglucosan (LG) and other monomeric anhydrosugars, furans, cyclopentanones, pyrans and other related derivatives. The ring scission process mainly obtains hydroxyacetaldehyde (HAA), acetol (HA), other linear carbonyls, linear alcohols, esters, and other products (Piskorz et al., 1986; Radlein et al., 1991; Lanza et al., 2009; Lin et al., 2009). In a recent study, Shen & Gu (2009) proposed the detailed possible routes for the pyrolytic formation of several major products from cellulose, as shown in Fig.2.

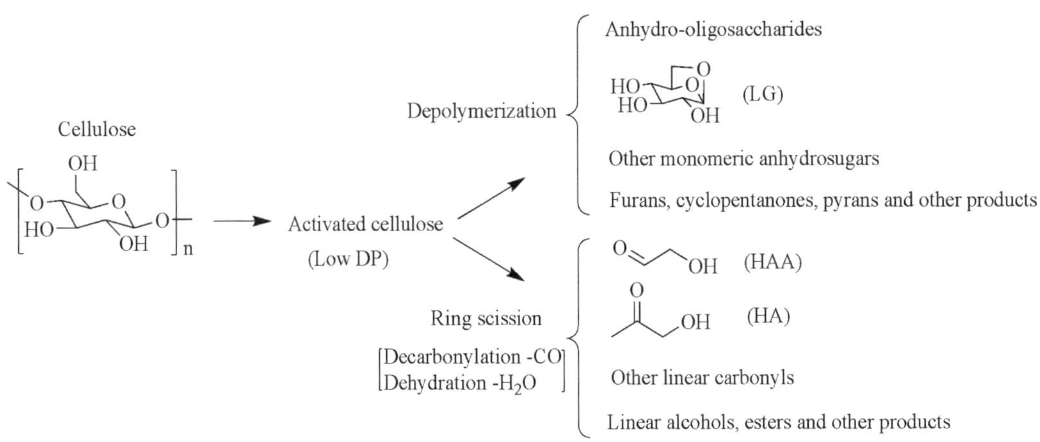


Fig. 1. The two parallel pyrolytic pathways during fast pyrolysis of cellulose at moderate temperatures

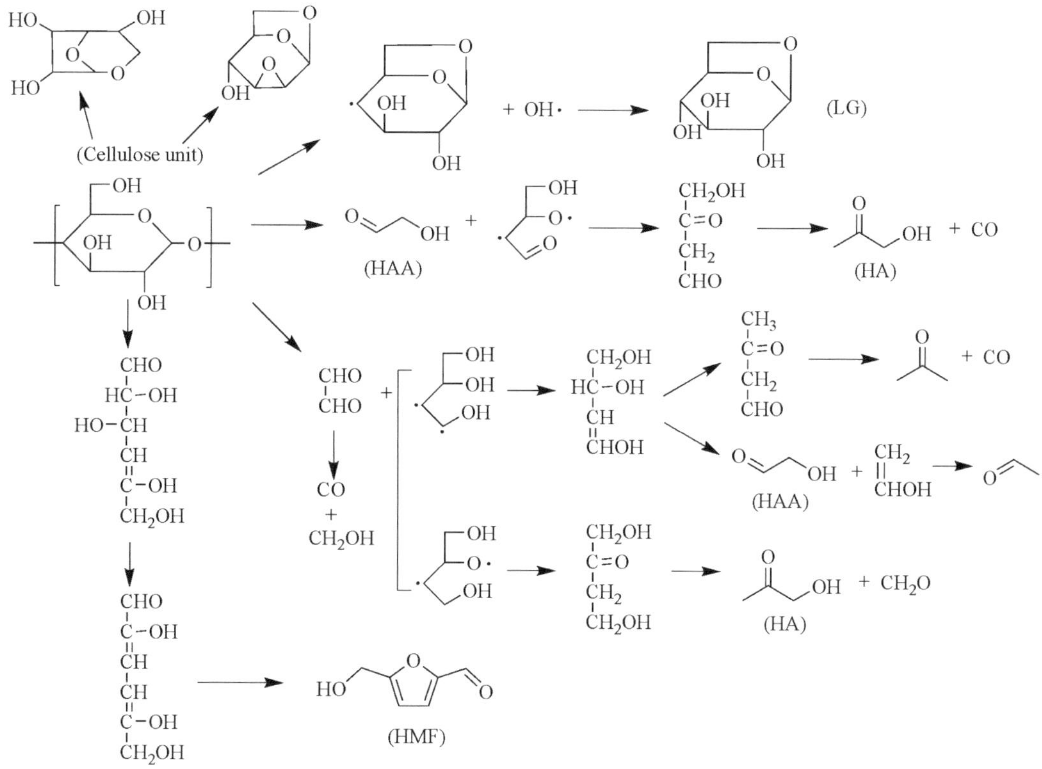

Fig. 2. The major pyrolytic pathways during cellulose fast pyrolysis (proposed by Shen & Gu)

## 2.2 Hemicellulose fast pyrolysis

Hemicelluloses are closely associated with cellulose in the cell wall as well as to lignin in the middle lamella. They are amorphous polysaccharides with building units belong either to hexoses (mainly D-glucose, D-mannose, and D-galactose) or to pentoses (mainly D-xylose and L-arabinose). The primary hemicellulose components are galactoglucomannans (glucomannans) and arabinoglucuronoxylan (xylan). Compared with cellulose, hemicelluloses have received less attention in their pyrolytic mechanism study.

Hemicelluloses are less thermally stable than cellulose, presumably due to the lack of crystallinity, and their pyrolysis is generally thought to be analogous to cellulose in the reaction mechanisms. Fast pyrolysis of glucomannans generates similar pyrolytic products as the cellulose, while the pyrolytic products from xylan differ more than those from cellulose (Alen et al., 1996). Generally, xylan fast pyrolysis will obtain higher char yield than that from cellulose, and will not form a typical depolymerization product, like the LG from cellulose. A probable cause of this difference was explained by Ponder & Nichards (1991). During cellulose pyrolysis, the glucosyl cation from the scission of the common glucans can readily form a stable 1,6-anhydride with the free primary hydroxyl group at C-6, and finally yields the volatile LG. However, there is no feasible mechanism for intramolecular "stabilization" for the xylosyl cation via anhydride formation, and hence, the xylosyl cation is more likely to enter the non-specific dehydration pathways, resulting in the char formation rather volatiles.

In a recent study, Shen et al. (2010) also proposed the detailed possible routes for the pyrolytic formation of several major products from main xylan chain, as well as the O-acetylxylan and 4-O-methylglucuronic acid unit, as shown in Fig.3.

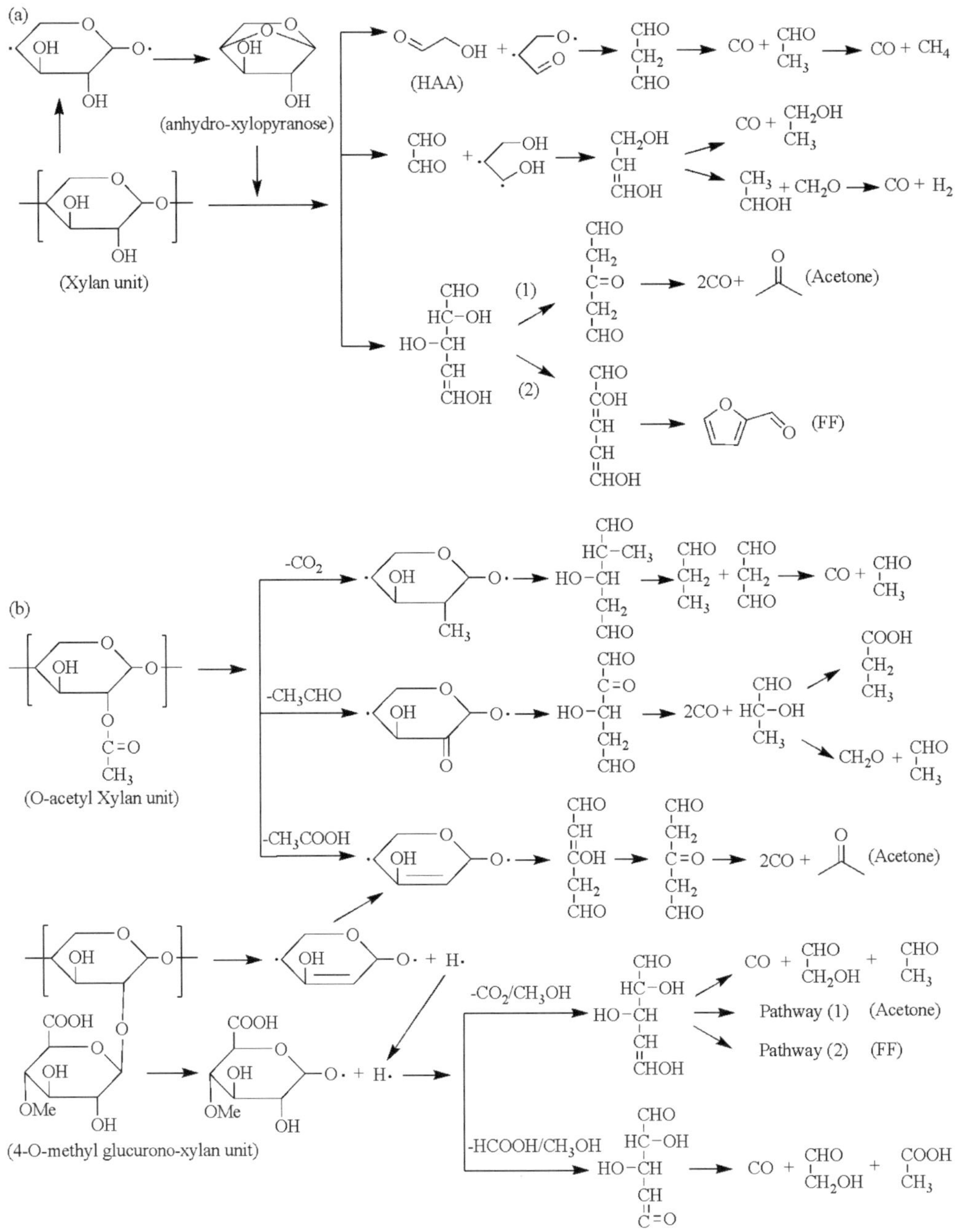


Fig. 3. The major pyrolytic pathways during xylan fast pyrolysis (proposed by Shen et al.) (a) the main chain of O-acetyl-4-methylglucurono-xylan, (b) O-acetylxylan and 4-O-methylglucuronic acid unit

### 2.3 Lignin fast pyrolysis

Lignin is the amorphous material that surrounds cellulose fibres and cements them together. It is a complex, heterogeneous polymer formed by the polymerization of three phenyl propane monomers, i.e. guaiacyl (4-hydroxy-3-methoxyphenyl), syringyl (3,5-dimethoxy-4-hydroxyphenyl) and *p*-hydroxyphenyl units. Lignin is the most complicated, least understood and most thermally stable component of biomass.

Primary pyrolysis of lignin begins with thermal softening at around 200°C, while most lignin pyrolysis occurs at higher temperatures, higher than the fast decomposition of cellulose. Fast pyrolysis of lignin will obtain higher char yield and lower liquid yield than holocellulose, and the liquid product can be classified into three groups, the large molecular oligomers (known as pyrolytic lignins), the monomeric phenolic compounds, as well as light compounds (such as methanol, HAA, acetic acid). The pyrolytic lignins are formed in much higher yield than the other two classes of products, usually account for 13.5-27.7 wt% of crude bio-oils on a water-free basis (Oasmaa et al., 2003; Garcia-Perez et al., 2007; Garcia-Perez et al., 2008). Several studies have been conducted to analyze them by various wet chemical and spectroscopic methods as well as pyrolysis-gas chromatography/mass spectrometry (Scholze & Meier, 2001; Scholze et al., 2001; Bayerbach et al., 2006; Garcia-Perez et al., 2008; Bayerbach & Meier, 2009), and to find out that their average molecular weight is between 650 and 1300 g/mol, and they are typically characterized by biphenyl, phenyl coumaran, diphenyl ethers, stilbene and resinol structures.

### 2.4 Biomass fast pyrolysis

Biomass fast pyrolysis is the summation of its major components fast pyrolysis. In addition to the cellulose, hemicellulose and lignin, biomass usually contains some extractives which would also decompose during fast pyrolysis, making the pyrolytic products more complex.

The biomass fast pyrolytic pathways and subsequent product distribution will be influenced by many factors, including biomass composition, feedstock property, pyrolysis temperature, heating rate, pressure, pyrolysis reactor configuration and a combination of these variables. The detailed effects of these factors can be found in previous studies, and will not be shown here.

## 3. Chemicals production from selective fast pyrolysis of biomass

### 3.1 Levoglucosan

LG (1,6-anhydro-β-D-glucopyranose) is the most important pyrolytic product of pure cellulose, formed through the depolymerization reaction. The chemistry of LG has long been known, and it can be used as a chiral synthon for the synthesis of stereoregular polysaccharides possessing biological activities (Miftakhov et al., 2001; Bailliez et al., 2004). Moreover, LG can also be hydrolyzed to glucose, providing a potentially rapid route to produce bio-ethanol (Bennett et al., 2009).

Fast pyrolysis of pure cellulose can produce LG with the yield up to 40 wt%, but fast pyrolysis of raw biomass materials would produce much lower LG yield due to the presence of inorganic impurities. Even the minor amounts of alkaline cations would shift the pyrolytic pathways of cellulose, to promote the formation of ring-scission products (such as HAA, HA, etc) and char on the expense of LG (Pan & Richards, 1989; NikAzar et al., 1997). Hence, it is necessary to use pure cellulose or demineralized biomass for the LG production. Furthermore, some studies pointed out that when small amounts of acids or acidic salts were added to demineralized biomass, the LG yield could be increased. For example, it was reported in an analytical pyrolysis study that, the LG content in the pyrolytic products was 5.3% from raw birch wood, 17.0% from decationized wood, 33.6% from wood impregnated with 1.0% phosphoric acid, and 27.3% from decationized wood adsorbed with iron ions (Dobele et al., 2005).

The main difficulty in LG preparation is not the pyrolysis process, but its isolation from pyrolysis liquids. Due to its high boiling point (386 °C), LG could not be simply recovered by distillation. Currently, several methods have been proposed or patented for the purification of LG (Howard et al., 1993; Moens, 1994; S.Scott et al., 1996).

### 3.2 Levoglucosenone

Levoglucosenone (LGO, 1,6-anhydro-3,4-dideoxy-β-D-pyranosen-2-one, or 6,8-dioxabicyclo[3.2.1]oct-2-en-4-one) is a sugar enone product of cellulose, formed from the combined depolymerization and dehydration reactions, with the possible pyrolytic pathways shown in Fig.4 (Ohnishi et al., 1975). Its structure was firstly confirmed in 1973 (Halpern et al., 1973), and further confirmed and adopted by other researchers. LGO is an optically active organic compound in which all carbon atoms are different environments and which has easily modifiable functional groups. As a result, LGO can be used in the synthesis of various products (such as tetrodotoxin, thiosugar, ras activation inhibitors). The detailed application ways of the LGO can be found elsewhere (Miftakhov et al., 1994).

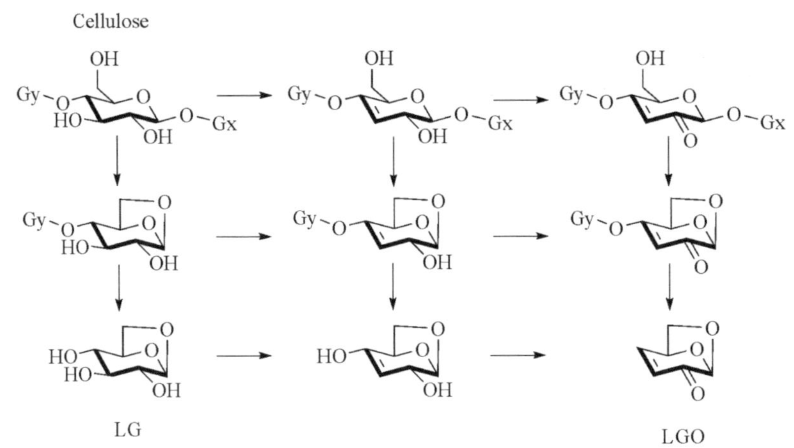


Fig. 4. The pyrolytic pathways for the formation of LGO from pyrolysis of cellulose

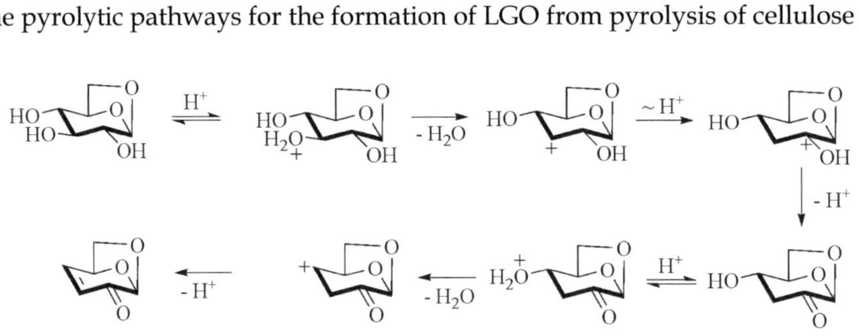


Fig. 5. The formation of LGO from acid-catalyzed decomposition of LG

LGO is formed in very low yield from fast pyrolysis of cellulose or biomass, but can be promoted by the addition of some acid catalysts in the pyrolytic process. A mechanism has been proposed for acid-catalyzed decomposition of LG to form the LGO, as shown in Fig.5 (Halpern et al., 1973).

Various acid catalysts exhibited the capability to promote the LGO formation, such as the $MgCl_2$ and $FeCl_3$ (Klampfl et al., 2006), $(NH_4)_2SO_4$ and $(NH_4)_2HPO_4$ (Pappa et al., 2006; Di Blasi et al., 2007), $CrO_3$ and $CrO_3$+$CuSO_4$ (Fu et al., 2008a), $ZnCl_2$ (Di Blasi et al., 2008), M/MCM-41(M=Sn, Zr, Ti, Mg, etc.) (Torri et al., 2009a). However, almost all of these catalysts did not show high selectivity on the LGO, because they catalyzed the formation of several dehydrated products (LGO, LAC, DGP, FF, etc.), rather than LGO alone.

According to a series of studies performed by Dobele et al. (1999, 2001, 2003), fast pyrolysis of cellulose/biomass impregnated with phosphoric acid could produce LGO with very high purity. The highest LGO yield reached 34 wt% from microcrystalline cellulose impregnated with 2% phosphoric acid, or 17.5 wt% from birch wood impregnated with 2.5% phosphoric acid. Other studies also confirmed the promising catalytic selectivity of the phosphoric acid on the LGO production (Sarotti et al., 2007; Fu et al., 2008b; Nowakowski et al., 2008). Furthermore, Dobele et al. (2005) reported that the pretreatment of cellulose/biomass with adsorption of $Fe_2(SO_4)_3$ provided another way to prepare LGO with high purity, but the selectivity of the $Fe_2(SO_4)_3$ on the LGO was a little lower than the phosphoric acid.

In a recent study, it was reported that fast pyrolysis of pure cellulose followed with catalytic cracking of the pyrolysis vapors with solid super acids (sulfated metal oxides, $SO_4^{2-}/TiO_2$, $SO_4^{2-}/ZrO_2$, $SO_4^{2-}/SnO_2$, etc.) allowed the production of LGO with the content reaching 40% (peak area% on the GC/MS ion chromatograms) in the pyrolytic products (Lu et al., 2009a). In fact, when the solid super acids were mechanically mixed with cellulose, fast pyrolysis of the mixture also produced LGO with high purity. Compared with the impregnation of catalysts (phosphoric acid or $Fe_2(SO_4)_3$) on the cellulose/biomass, the utilization of solid catalysts avoids the complex pretreatment process, and will offer a significant advantage on catalyst recycles.

Compared with the LG, the LGO can be easily recovered from pyrolytic liquids by distillation, and a detailed purification method was proposed by Marshall (2008).

### 3.3 1-hydroxy-3,6-dioxabicyclo[3.2.1]octan-2-one

The 1-hydroxy-3,6-dioxabicyclo[3.2.1]octan-2-one (LAC) is another multifunctional C6-monomer formed from fast pyrolysis of cellulose. It was firstly identified by Fruneaxu et al. (1988), who also proposed its possible pyrolytic formation pathways, as shown in Fig.6. Compared with the LG and LGO, the LAC has received much less attention in either its production or application studies.

Similar as the LGO, the LAC is formed in very low yield during fast pyrolysis of cellulose, but can be promoted in the acid-catalyzed pyrolysis process. Most of the catalysts which could increase the LGO formation, could also promote the LAC formation. Therefore, it is necessary to find out proper catalyst which has high selectivity on the LAC.

According to some previous studies, fast pyrolysis of cellulose mixed with nanopowder aluminium titanate at 350 °C obtained the LGO and LAC as the major products, with the yields reaching 22 wt% and 8.6 wt%, respectively. When the pyrolysis temperature was increased to 500 °C, the LAC was formed as the only predominant product with high purity, but its yield was decreased to 6.2 wt% (the LGO yield was decreased to 0.77 wt%) (Fabbri et

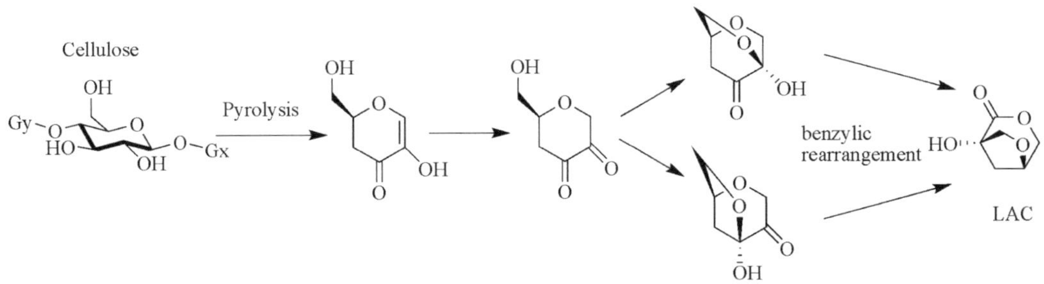


Fig. 6. The pyrolytic pathways for the formation of LAC from pyrolysis of cellulose

al., 2007a; Fabbri et al., 2007b; Torri et al., 2009b). As been indicated above, the utilization of solid catalysts will provide a convenient way for the catalyst recycles. Further studies have also been conducted by them to use the LAC as a possible building block in the synthesis of fine chemicals.

### 3.4 Anhydro-oligosaccharides

It is clearly demonstrated in previous studies that fast pyrolysis of cellulose will generate a range of anhydro-oligosaccharides, resulting from random cleavage of the polymer chain (Radlein et al., 1987; Pouwels et al., 1989; Lomax et al., 1991; Radlein et al., 1991). Anhydro-oligosaccharides are potential for a number of possible uses, such as the preparation of glycoconjugates, the so-called anti-adhesive drugs, and so on.

Although fast pyrolysis is a known technique to produce anhydro-oligosaccharides from cellulose, very limited studies have been carried out to produce the oligomers as a target product. Piskorz et al. (2000) initiated a flash pyrolysis study on this target product. Compared with conventional fast pyrolysis aiming at the maximum bio-oil yield, the production of anhydro-oligosaccharides required stricter reaction conditions: higher pyrolysis temperature (850~1200 °C) together with shorter residence time (35~75 ms). The short residence time was used to inhibit the conversion of large oligomers to monomer and dimer anhydrosaccharides. Anhydro-oligosaccharides in the range of G2-G7 were successfully produced with the yield up to near 20 wt%. These oligomers could be recovered as water soluble fraction from the reaction solid residues. Moreover, substantial amounts of larger oligomers (> G7) could also be produced but difficult to be identified.

Furthermore, if single anhydro-oligosaccharides can be produced and recovered, they should be more valuable than the mixed anhydro-oligosaccharides, but no reports are available in this research field at present.

### 3.5 Furfural

Furfural (FF) is a typical pyrolytic product formed from both of cellulose and hemicellulose. It is widely used as an organic solvent or an organic reagent for the production of medicines, resins, food additives, fuel additives and other special chemicals. Currently, FF is industrially produced from agricultural raw materials rich in pentosan. By aqueous acid catalysis (e.g., sulfuric acid or phosphoric acid), the pentosan is firstly hydrolyzed to pentose which is then dehydrated to form FF (Zeitsch, 2000).

Similar as the LGO and LAC, the formation of FF can be promoted in the acid-catalyzed pyrolysis of biomass (Shafizadeh & Lai, 1972; Encinar et al., 1997; Encinar et al., 1998; Adam et al., 2005; Shimada et al., 2008; Lu et al., 2009b). Zinc chloride ($ZnCl_2$) is one of the promising catalysts to produce FF. It was found that fast pyrolysis of cellulose impregnated with small amounts of $ZnCl_2$ (below 10 wt%) would generate several dehydrated products as the primary pyrolytic products, mainly the FF, LGO, LAC and 1,4:3,6-dianhydro-α-D-glucopyranose (DGP). With the increasing of $ZnCl_2$ impregnation (at least 15 wt% or more), the $ZnCl_2$ catalysis would increase the FF formation, while decrease the anhydrosugars. Moreover, the secondary catalytic cracking of the primary pyrolysis vapors by $ZnCl_2$ could promote the conversion of LGO and other anhydrosugars to FF, leaving the FF as the only predominant product (Fig.7). Compared with the cellulose, the $ZnCl_2$-catalzyed fast pyrolysis of xylan would obtain the FF as the only predominant product, and the FF yield would be higher than that from cellulose, indicating that biomass rich xylan would be suitable for the FF production.

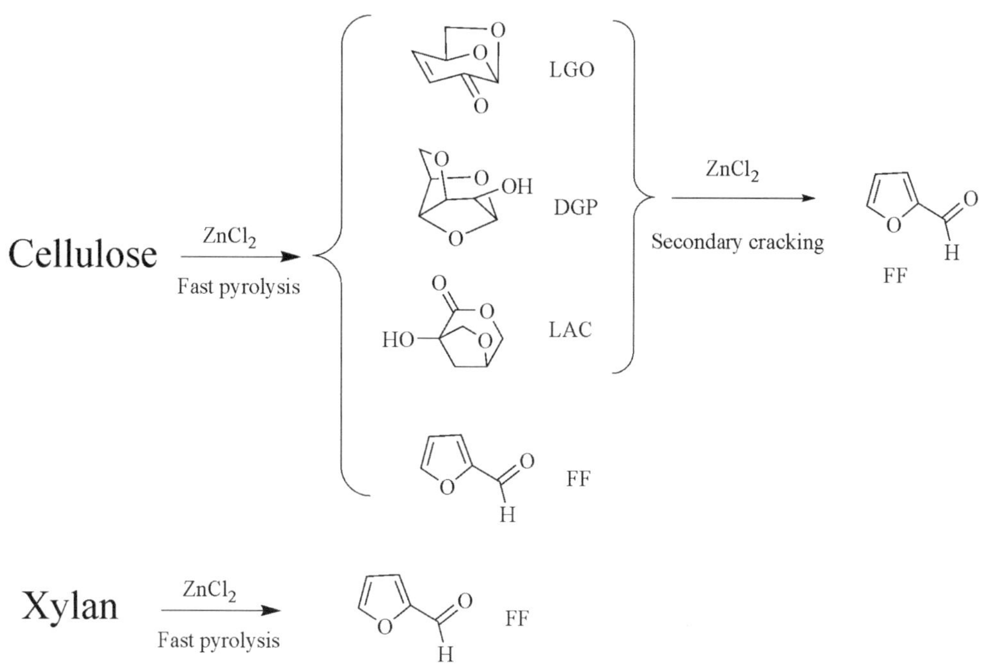


Fig. 7. The formation of FF during $ZnCl_2$-catalyzed fast pyrolysis of biomass

In another study, microwave-assisted fast pyrolysis was applied to treat biomass mixed with various catalysts, and the results revealed that the $MgCl_2$ exhibited high selectivity on the FF production (Wan et al., 2009). The highest FF content was more than 80% (peak area% on the GC/MS ion chromatograms) at the 8 g $MgCl_2$ mixed with 100 g biomass.

In addition, $ZnCl_2$ (or sometimes $MgCl_2$) is an effective chemical activation agent for the production of activated carbons from biomass. Hence, for the biomass pretreated by $ZnCl_2$ (or $MgCl_2$) impregnation, they can be firstly subjected to fast pyrolysis to produce FF. The solid residues which contained char and $ZnCl_2$ (or $MgCl_2$), could be further activated to produce activated carbons, so as to achieve the co-production of FF and activated carbons.

### 3.6 Hydroxyacetaldehyde

HAA is formed as the most abundant linear carbonyl product from pyrolytic ring scission of holocellulose. It is the simplest aldehyde-alcohol or sugar, and can be used as an active meat-browning agent in food flavoring industry (Garham & Underwood, 1993), or ingredient in cosmetic industry.

As been indicated above, during fast pyrolysis of biomass, the pyrolytic ring scission of holocellulose will be promoted by the small amounts of ash or alkaline cations (Shimada et al., 2008), and hence, HAA can be formed with high yield in these conditions. Moreover, HAA could be further formed from the secondary cracking of the anhydrosugar products. As a result, HAA is usually the most abundant single organic compound in crude bio-oils produced from raw biomass materials.

Similar as the LG, the difficulty for the preparation of HAA mainly lies in its recovery from pyrolysis liquids, due to its high reactivity. An efficient purification method was patented by Stradal & Underwood (1995).

### 3.7 Acetic acid

During fast pyrolysis of biomass, acetic acid (AA) can be formed from different ways, with a large portion from deacetylation of hemicellulose, and a small portion from pyrolytic ring scission of holocellulose and side-chain cracking of lignin. The content of AA could be over 10 wt% in crude bio-oils produced from biomass rich in acetyl groups, and some preliminary studies have been reported for the recovery of AA (Mahfud et al., 2008). AA is a common chemical, and is less valuable than other chemicals described above. According to its formation characteristics, selective fast pyrolysis of biomass to produce AA with high purity might be achieved in several different ways.

The deacetylation of hemicellulose occurs at relatively low temperature, lower than the fast decomposition of the major biomass components. Hence, AA can be produced from low temperature pyrolysis of biomass. It has already been confirmed that during carbonization or torrefaction of biomass at temperatures lower than 300 °C, a liquid by-product rich in AA could be produced (Prins et al., 2006). In fact, if biomass is subjected to fast pyrolysis at around 300 °C for a short time period, part of the acetyl group would break down to form AA, while the other biomass components would not be greatly influenced. Therefore, a two-step process might be applied to treat biomass, with the first low-temperature fast pyrolysis to produce AA, followed with the moderate-temperature fast pyrolysis to produce bio-oil.

AA can also be prepared by catalytic pyrolysis of biomass. For example, Qi et al. (2006) reported that catalytic pyrolysis of biomass using NaY catalyst produced abundant AA, with the content over 80% (peak area% on the GC/MS ion chromatograms) in the pyrolytic products.

In addition, during the production of FF through fast pyrolysis of biomass impregnated with $ZnCl_2$ (or $MgCl_2$), the catalysis of $ZnCl_2$ (or $MgCl_2$) inhibited most of the pyrolytic pathways, but not the AA formation. As a result, AA would be the second most important product in the catalytic organic products, and thus, it could be recovered as a byproduct during preparation of FF.

### 3.8 Phenolic compounds

The lignin-derived phenolic compounds (monomeric phenolic compounds and pyrolytic lignins) can be isolated from crude bio-oils by different methods (Achladas, 1991; Amen-

Chen et al., 1997; Deng et al., 2009), and they are known to be used as phenol replacement in production of phenol-formaldehyde resins (Himmelblau, 1991; Chum & Kreibich, 1992; Roy et al., 2000; Giroux et al., 2001).

As indicated above, the monomeric phenolic compounds are usually formed in much lower yield than the pyrolytic lignins. Since the monomers are more reactive than the oligomers for resin production, and thus, it will be attractive to promote the production of monomeric phenolic compounds. According to some previous studies, fast pyrolysis of biomass impregnated with some alkaline compounds (NaOH, KOH) could increase the yield of monomeric phenolic compounds (Nowakowski et al., 2007; Di Blasi et al., 2009a; Di Blasi et al., 2009b). In another study, Pd/SBA-15 catalysts were employed to catalytic crack biomass fast pyrolysis vapors that contained a lot of pyrolytic lignins, and the results indicated that the Pd/SBA-15 catalysts were able to promote the conversion the pyrolytic lignins to monomeric phenolic compounds, and meanwhile to crack and decrease holocellulose-derived products. Hence, the content of the total monomeric phenolic compounds reached 55% (peak area% on the GC/MS ion chromatograms) in the catalytic pyrolytic products (Lu et al., 2010a). Some other catalysts were also confirmed to possess the catalytic capability to increase the yield of monomeric phenolic compounds or their content in the catalytic bio-oils (Adam et al., 2006; Antonakou et al., 2006; Nilsen et al., 2007; Ates & Isikdag, 2009).

Furthermore, if single phenolic compounds could be produced and recovered, they should be much more valuable than the mixed phenolic compounds. According to a previous study done by Murwanashyaka et al. (2001a), the evolution of major monomeric phenolic compounds took place in the following order: methylguaiacol, ethylguaiacol, guaiacol, propenylsyringol, phenol and catechol, which suggested that the production of specific phenolic compounds might be achieved by step-wise pyrolysis of biomass. Catalytic pyrolysis is another promising way. For example, during the catalytic cracking of biomass fast pyrolysis vapors with Pd/SBA-15 catalysts, the 4-ethyl-2-methoxy-phenol was increased greatly, with the content up to 10% (peak area% on the GC/MS ion chromatograms) in the pyrolytic products (Lu et al., 2010a). Some subsequent studies have also been reported for the recovery of pure single phenolic compounds (Murwanashyaka et al., 2001b).

### 3.9 Light aromatic hydrocarbons

Hydrocarbons are usually formed in very low yields during fast pyrolysis of biomass, but can be greatly increased by using proper cracking catalysts with deoxygenation capability (Bridgwater, 1996). Zeolite catalysts (such as HZSM-5, HY, etc.) are very effective to convert the highly oxygenated crude bio-oils or pyrolysis vapors to hydrocarbons which are dominated by several light aromatic hydrocarbons (benzene, toluene, xylene and naphthalene) (Horne & Williams, 1996; Olazar et al., 2000; Williams & Nugranad, 2000; Pattiya et al., 2008; Zhang et al., 2009). For example, in the studies performed by Adjaye et al. (1995, 1996), catalytic cracking of the crude bio-oil by HZSM-5 catalysts obtained a organic liquid product with up to 90 wt% of aromatic hydrocarbons, and the aromatic hydrocarbons contained abundant toluene (31.8 wt%) and xylene (33.1 wt%).

Other catalysts were also investigated for the production of light aromatic hydrocarbons. For example, Wang et al. (2008) reported that catalytic pyrolysis of biomass using CoMo-S/$Al_2O_3$ catalyst produced the four light aromatic hydrocarbons (benzene, toluene, xylene and naphthalene) with the yield reaching 6.3 wt% at 590 °C.

### 3.10 Other valuable chemicals

In addition to the above chemicals, many other chemicals can be produced by selective fast pyrolysis of biomass. For example, Chen et al. (2008) reported that fast pyrolysis of biomass impregnated with $Na_2CO_3$ produced HA with high purity. Badri et al. (2008) revealed that catalytic pyrolysis of cotton with reactive dyes favored the formation of 5-hydroxymethyl-furfural (HMF) and 3-(hydroxymethyl)-furan. Lu et al. (2010a) found that fast pyrolysis of cellulose followed with catalytic cracking of the vapors by sulfated metal oxides could obtain high yields of furan and 5-methyl furfural. In another study, Lu et al. (2010b) confirmed that catalytic cracking of the biomass fast pyrolysis vapors using $ZrO_2$&$TiO_2$ increased the formation of three light carbonyl products (acetaldehyde, acetone and 2-butanone).

## 4. Conclusion

Most of the selective fast pyrolysis techniques are only in their early stage of development, and none of the techniques is commercially practical to produce specific chemicals in marketable quantities at present. Three aspects should be considered for the commercialization of the selective fast pyrolysis techniques, including (1) the technique to produce specific chemicals in high yield and purity, (2) the method to recover the target chemicals from pyrolysis liquids, and (3) the ready markets for the chemicals.

Among the above indicated chemicals, the LG, HAA and AA can be produced without catalyst utilization, and thus, their large scale production might be relatively easy to achieve through slight modification of the conventional fast pyrolysis technique. While the production of other chemicals requires catalysts, which will add difficulty to their scale up. Various methods have been proposed for the chemicals recovery, and further studies should be conducted to reduce the purification cost. Finally, it is important to note that ready markets do not exist for the LG, LGO, LAC, anhydro-oligosaccharides, and some other chemicals at present. Corresponding markets should be developed by manufacturers who would incorporate these chemicals into various products.

## 5. References

Achladas, G.E. (1991). Analysis of biomass pyrolysis liquids - separation and characterization of phenols. *J. Chromatogr.*, 542, 263-275.

Adam, J.; Blazso, M.; Meszaros, E.; Stocker, M.; Nilsen, M.H.; Bouzga, A.; Hustad, J.E.; Gronli, M. & Oye, G. (2005). Pyrolysis of biomass in the presence of Al-MCM-41 type catalysts. *Fuel*, 84, 1494-1502.

Adam, J.; Antonakou, E.; Lappas, A.; Stocker, M.; Nilsen, M.H.; Bouzga, A.; Hustad, J.E. & Oye, G. (2006). In situ catalytic upgrading of biomass derived fast pyrolysis vapours in a fixed bed reactor using mesoporous materials. *Micropor. Mesopor. Mat.*, 96, 93-101.

Adjaye, J.D. & Bakhshi, N.N. (1995). Production of hydrocarbons by catalytic upgrading of a fast pyrolysis bio-oil .1. Conversion over various catalysts. *Fuel Process. Technol.*, 45, 161-183.

Adjaye, J.D.; Katikaneni, S.P.R. & Bakhshi, N.N. (1996). Catalytic conversion of a biofuel to hydrocarbons: effect of mixtures of HZSM-5 and silica-alumina catalysts on product distribution. *Fuel Process. Technol.*, 48, 115-143.

Alen, R.; Kuoppala, E. & Oesch, P. (1996). Formation of the main degradation compound groups from wood and its components during pyrolysis. *J. Anal. Appl. Pyrol.*, 36, 137-148.

Amen-Chen, C.; Pakdel, H. & Roy, C. (1997). Separation of phenols from Eucalyptus wood tar. *Biomass & Bioenergy*, 13, 25-37.

Antonakou, E.; Lappas, A.; Nilsen, M.H.; Bouzga, A. & Stocker, M. (2006). Evaluation of various types of Al-MCM-41 materials as catalysts in biomass pyrolysis for the production of bio-fuels and chemicals. *Fuel*, 85, 2202-2212.

Ates, F. & Isikdag, M.A. (2009). Influence of temperature and alumina catalyst on pyrolysis of corncob. *Fuel*, 88, 1991-1997.

Badri, B. (2008). The influence of reactive dyes on the pyrolysis of cotton. *J. Anal. Appl. Pyrol.*, 81, 162-166.

Bailliez, V.; Olesker, A. & Cleophax, J. (2004). Synthesis of polynitrogenated analogues of glucopyranoses from levoglucosan. *Tetrahedron*, 60, 1079-1085.

Bayerbach, R.; Nguyen, V.D.; Schurr, U. & Meier, D. (2006). Characterization of the water-insoluble fraction from fast pyrolysis liquids (pyrolytic lignin) - Part III. Molar mass characteristics by SEC, MALDI-TOF-MS, LDI-TOF-MS, and Py-FIMS. *J. Anal. Appl. Pyrol.*, 77, 95-101.

Bayerbach, R. & Meier, D. (2009). Characterization of the water-insoluble fraction from fast pyrolysis liquids (pyrolytic lignin). Part IV: Structure elucidation of oligomeric molecules. *J. Anal. Appl. Pyrol.*, 85, 98-107.

Bennett, N.M.; Helle, S.S. & Duff, S.J.B. (2009). Extraction and hydrolysis of levoglucosan from pyrolysis oil. *Bioresour. Technol.*, 100, 6059-6063.

Boutin, O.; Ferrer, M. & Lede, J. (1998). Radiant flash pyrolysis of cellulose - Evidence for the formation of short life time intermediate liquid species. *J. Anal. Appl. Pyrol.*, 47, 13-31.

Bridgwater, A. (1996). Production of high grade fuels and chemicals from catalytic pyrolysis of biomass. *Catal. Today*, 29, 285-295.

Bridgwater, A.V. & Peacocke, G.V.C. (2000). Fast pyrolysis processes for biomass. *Renew. Sust. Energ. Rev.*, 4, 1-73.

Chen, M.Q.; Wang, J.; Zhang, M.X.; Chen, M.G.; Zhu, X.F.; Min, F.F. & Tan, Z.C. (2008). Catalytic effects of eight inorganic additives on pyrolysis of pine wood sawdust by microwave heating. *J. Anal. Appl. Pyrol.*, 82, 145-150.

Chum, H.L. & Kreibich, R.E., 1992. Process for preparing phenolic formaldehyde resole resin products derived from fractionated fast-pyrolysis oils. US5091499.

Deng, L.; Yan, Z.; Fu, Y. & Guo, Q.X. (2009). Green solvent for flash pyrolysis oil separation. *Energ Fuel*, 23, 3337-3338.

Di Blasi, C.; Branca, C. & Galgano, A. (2007). Effects of diammonium phosphate on the yields and composition of products from wood pyrolysis. *Ind. Eng. Chem. Res.*, 46, 430-438.

Di Blasi, C.; Branca, C. & Galgano, A. (2008). Products and global weight loss rates of wood decomposition catalyzed by zinc chloride. *Energ Fuel*, 22, 663-670.

Di Blasi, C.; Galgano, A. & Branca, C. (2009a). Effects of potassium hydroxide impregnation on wood pyrolysis. *Energ Fuel*, 23, 1045-1054.

Di Blasi, C.; Galgano, A. & Branca, C. (2009b). Influences of the chemical state of alkaline compounds and the nature of alkali metal on wood pyrolysis. *Ind. Eng. Chem. Res.*, 48, 3359-3369.

Dobele, G.; Rossinskaja, G.; Telysheva, G.; Meier, D. & Faix, O. (1999). Cellulose dehydration and depolymerization reactions during pyrolysis in the presence of phosphoric acid. *J. Anal. Appl. Pyrol.*, 49, 307-317.

Dobele, G.; Meier, D.; Faix, O.; Radtke, S.; Rossinskaja, G. & Telysheva, G. (2001). Volatile products of catalytic flash pyrolysis of celluloses. *J. Anal. Appl. Pyrol.*, 58, 453-463.

Dobele, G.; Dizhbite, T.; Rossinskaja, G.; Telysheva, G.; Mier, D.; Radtke, S. & Faix, O. (2003). Pre-treatment of biomass with phosphoric acid prior to fast pyrolysis - A promising method for obtaining 1,6-anhydrosaccharides in high yields. *J. Anal. Appl. Pyrol.*, 68-9, 197-211.

Dobele, G.; Rossinskaja, G.; Diazbite, T.; Telysheva, G.; Meier, D. & Faix, O. (2005). Application of catalysts for obtaining 1,6-anhydrosaccharides from cellulose and wood by fast pyrolysis. *J. Anal. Appl. Pyrol.*, 74, 401-405.

Encinar, J.M.; Beltran, F.J.; Ramiro, A. & Gonzalez, J.F. (1997). Catalyzed pyrolysis of grape and olive bagasse. Influence of catalyst type and chemical treatment. *Ind. Eng. Chem. Res.*, 36, 4176-4183.

Encinar, J.M.; Beltran, F.J. & Ramiro, A. (1998). Pyrolysis/gasification of agricultural residues by carbon dioxide in the presence of different additives: influence of variables. *Fuel Process. Technol.*, 55, 219-233.

Evans, R.J. & Milne, T.A. (1987). Molecular characterization of the pyrolysis of biomass. 1. Fundamentals. *Energ Fuel*, 1, 123-137.

Fabbri, D.; Torri, C. & Baravelli, V. (2007a). Effect of zeolites and nanopowder metal oxides on the distribution of chiral anhydrosugars evolved from pyrolysis of cellulose: An analytical study. *J. Anal. Appl. Pyrol.*, 80, 24-29.

Fabbri, D.; Torri, C. & Mancini, I. (2007b). Pyrolysis of cellulose catalysed by nanopowder metal oxides: Production and characterisation of a chiral hydroxylactone and its role as building block. *Green Chem*, 9, 1374-1379.

Fu, Q.R.; Argyropoulos, D.S.; Tilotta, D.C. & Lucia, L.A. (2008a). Understanding the pyrolysis of CCA-treated wood Part I. Effect of metal ions. *J. Anal. Appl. Pyrol.*, 81, 60-64.

Fu, Q.R.; Argyropoulos, D.S.; Tilotta, D.C. & Lucia, L.A. (2008b). Understanding the pyrolysis of CCA-treated wood. Part II. Effect of phosphoric acid. *J. Anal. Appl. Pyrol.*, 82, 140-144.

Furneaux, R.H.; M.Mason, J. & J.Miller, I. (1988). A novel hydroxylactone from the Lewis acid catalyzed pyrolysis of cellulose. *J. Chem. Soc., Perkin Trans. 1*, 49-51.

Garcia-Perez, M.; Chaala, A.; Pakdel, H.; Kretschmer, D. & Roy, C. (2007). Characterization of bio-oils in chemical families. *Biomass & Bioenergy*, 31, 222-242.

Garcia-Perez, M.; Wang, S.; Shen, J.; Rhodes, M.; Lee, W.J. & Li, C.Z. (2008). Effects of temperature on the formation of lignin-derived oligomers during the fast pyrolysis of Mallee woody biomass. *Energ Fuel*, 22, 2022-2032.

Garham, R.G. & Underwood, G.L., 1993. Method of using fast pyrolysis liquids as liquid smoke. WO9105484.

Giroux, R.; Freel, B. & Graham, R., 2001. Novel natural resin formulations. US6326461.

Halpern, Y.; Riffer, R. & Broido, A. (1973). Levoglucosenone (1,6-Anhydro-3,4-dideoxy-Δ3-β-D-Pyranosen-2-one). A major product of the acid-catalyzed pyrolysis of cellulose and related carbohydrates. *J. Org. Chem.*, 38, 204-209.

Himmelblau, A., 1991. Method and apparatus for producing water-soluble resin and resin product made by that method. US5034498.

Horne, P.A. & Williams, P.T. (1996). Upgrading of biomass-derived pyrolytic vapours over zeolite ZSM-5 catalyst: effect of catalyst dilution on product yields. *Fuel*, 75, 1043-1050.

Howard, J.; Longley, C.; Morrison, A. & Fung, D. (1993). Process for isolating levoglucosan from pyrolysis liquids. CA2084906.

Klampfl, C.W.; Breuer, G.; Schwarzinger, C. & Koll, B. (2006). Investigations on the effect of metal ions on the products obtained from the pyrolysis of cellulose. *Acta Chim. Slov.*, 53, 437-443.

Lanza, R.; Nogare, D.D. & Canu, P. (2009). Gas phase chemistry in cellulose fast pyrolysis. *Ind. Eng. Chem. Res.*, 48, 1391-1399.

Lin, Y.C.; Cho, J.; Tompsett, G.A.; Westmoreland, P.R. & Huber, G.W. (2009). Kinetics and mechanism of cellulose pyrolysis. *J. Phys. Chem. C*, 113, 20097-20107.

Lomax, J.A.; Commandeur, J.M.; Arisz, P.W. & Boon, J.J. (1991). Characterization of oligomers and sugar ring-cleavage products in the pyrolysate of cellulose. *J. Anal. Appl. Pyrol.*, 19, 65-79.

Lu, Q.; Xiong, W.M.; Li, W.Z.; Guo, Q.X. & Zhu, X.F. (2009a). Catalytic pyrolysis of cellulose with sulfated metal oxides: A promising method for obtaining high yield of light furan compounds. *Bioresour. Technol.*, 100, 4871-4876.

Lu, Q.; Li, W.Z.; Zhang, D. & Zhu, X.F. (2009b). Analytical pyrolysis-gas chromatography/mass spectrometry (Py-GC/MS) of sawdust with Al/SBA-15 catalysts. *J. Anal. Appl. Pyrol.*, 84, 131-138.

Lu, Q.; Tang, Z.; Zhang, Y. & Zhu, X.F. (2010a). Catalytic Upgrading of Biomass Fast Pyrolysis Vapors with Pd/SBA-15 Catalysts. *Ind. Eng. Chem. Res.*, 49, 2573-2580.

Lu, Q.; Zhang, Y.; Tang, Z.; Li, W.Z. & Zhu, X.F. (2010b). Catalytic upgrading of biomass fast pyrolysis vapors with titania and zirconia/titania based catalysts. *Fuel*, doi: 10.1016/j.fuel.2010.02.030.

Mahfud, F.H.; van Geel, F.P.; Venderbosch, R.H. & Heeres, H.J. (2008). Acetic acid recovery from fast pyrolysis oil. An exploratory study on liquid-liquid reactive extraction using aliphatic tertiary amines. *Sep. Sci. Technol.*, 43, 3056-3074.

Marshall, J.A., 2008. An improved preparation of levoglucosenone from cellulose. Iowa State University, Master thesis.

Miftakhov, M.S.; Valeev, F.A. & Gaisina, I.N. (1994). Levoglucosenone: the properties, reactions, and use in fine organic synthesis. *Russ. Chem. Rev.*, 63, 869-882.

Miftakhov, M.S.; Ermolenko, M.S.; Gaisina, I.N.; Kuznetsov, O.M.; Selezneva, N.K.; Yusupova, Z.A. & Muslukhov, R.R. (2001). Chemistry of natural compounds, bioorganic, and biomolecular chemistry - Synthesis of a C(9)-C(13) fragment of acutiphycin from levoglucosan. *Russ. Chem. Bull.*, 50, 1101-1106.

Moens, L., 1994. Isolation of levoglucosan from pyrolysis oil derived from cellulose. WO9405704.

Mohan, D.; Pittman, C.U. & Steele, P.H. (2006). Pyrolysis of wood/biomass for bio-oil: A critical review. *Energ Fuel*, 20, 848-889.

Murwanashyaka, J.N.; Pakdel, H. & Roy, C. (2001a). Step-wise and one-step vacuum pyrolysis of birch-derived biomass to monitor the evolution of phenols. *J. Anal. Appl. Pyrol.*, 60, 219-231.

Murwanashyaka, J.N.; Pakdel, H. & Roy, C. (2001b). Seperation of syringol from birch wood-derived vacuum pyrolysis oil. *Sep. Purif. Technol.*, 24, 155-165.

NikAzar, M.; Hajaligol, M.R.; Sohrabi, M. & Dabir, B. (1997). Mineral matter effects in rapid pyrolysis of beech wood. *Fuel Process. Technol.*, 51, 7-17.

Nilsen, M.H.; Antonakou, E.; Bouzga, A.; Lappas, A.; Mathisen, K. & Stocker, M. (2007). Investigation of the effect of metal sites in Me-Al-MCM-41 (Me = Fe, Cu or Zn) on the catalytic behavior during the pyrolysis of wooden based biomass. *Micropor. Mesopor. Mat.*, 105, 189-203.

Nowakowski, D.J.; Jones, J.M.; Brydson, R.M.D. & Ross, A.B. (2007). Potassium catalysis in the pyrolysis behaviour of short rotation willow coppice. *Fuel*, 86, 2389-2402.

Nowakowski, D.J.; Woodbridge, C.R. & Jones, J.M. (2008). Phosphorus catalysis in the pyrolysis behaviour of biomass. *J. Anal. Appl. Pyrol.*, 83, 197-204.

Oasmaa, A.; Kuoppala, E. & Solantausta, Y. (2003). Fast pyrolysis of forestry residue. 2. Physicochemical composition of product liquid. *Energ Fuel*, 17, 433-443.

Ohnishi, A.; Kato, K. & Takagi, E. (1975). Curie-point pyrolysis of cellulose. *Polym. J.*, 7, 431-437.

Olazar, M.; Aguado, R.; Bilbao, J. & Barona, A. (2000). Pyrolysis of sawdust in a conical spouted-bed reactor with a HZSM-5 catalyst. *AIChE J.*, 46, 1025-1033.

Pan, W.P. & Richards, G.N. (1989). Influence of metal ions on volatile products of pyrolysis of wood. *J. Anal. Appl. Pyrol.*, 16, 117-126.

Pappa, A.; Mikedi, K.; Tzamtzis, N. & Statheropoulos, M. (2006). TG-MS analysis for studying the effects of fire retardants on the pyrolysis of pine-needles and their components. *J. Therm. Anal. Calorim.*, 84, 655-661.

Pattiya, A.; Titiloye, J.O. & Bridgwater, A.V. (2008). Fast pyrolysis of cassava rhizome in the presence of catalysts. *J. Anal. Appl. Pyrol.*, 81, 72-79.

Piskorz, J.; Radlein, D. & S.Scott, D. (1986). On the mechanism of the rapid pyrolysis of cellulose. *J. Anal. Appl. Pyrol.*, 9, 121-137.

Piskorz, J.; Radlein, D.; Scott, D.S. & Czernik, S. (1989). Pretreatment of wood and cellulose for production of sugars by fast pyrolysis. *J. Anal. Appl. Pyrol.*, 16, 127-142.

Piskorz, J.; Majerski, P.; Radlein, D.; Vladars-Usas, A. & Scott, D.S. (2000). Flash pyrolysis of cellulose for production of anhydro-oligomers. *J. Anal. Appl. Pyrol.*, 56, 145-166.

Ponder, G.R. & Richards, G.N. (1991). Thermal synthesis and pyrolysis of a xylan. *Carbohydr. Res.*, 218, 143-155.

Pouwels, A.D.; Eijkel, G.B.; Arisz, P.W. & Boon, J.J. (1989). Evidence for oligomers in pyrolysates of microcrystalline cellulose. *J. Anal. Appl. Pyrol.*, 15, 71-84.

Prins, M.J.; Ptasinski, K.J. & Janssen, F. (2006). Torrefaction of wood - Part 2. Analysis of products. *J. Anal. Appl. Pyrol.*, 77, 35-40.

Qi, W.Y.; Hu, C.W.; Li, G.Y.; Guo, L.H.; Yang, Y.; Luo, J.; Miao, X. & Du, Y. (2006). Catalytic pyrolysis of several kinds of bamboos over zeolite NaY. *Green Chem*, 8, 183-190.

Radlein, D.; Grinshpun, A.; Piskorz, J. & Scott, D.S. (1987). On the presence of anhydro-oligosaccharides in the sirups from the fast pyrolysis of cellulose. *J. Anal. Appl. Pyrol.*, 12, 39-49.

Radlein, D.; Piskorz, J. & Scott, D.S. (1991). Fast pyrolysis of natural polysaccharides as a potential Industrial-process. *J. Anal. Appl. Pyrol.*, 19, 41-63.

Roy, C.; Lu, X. & Pakdel, H., 2000. Process for the production of phenolic-rich pyrolysis oils for use in making phenol-formaldehyde resole resins. US6143856.

S.Scott, D.; Piskorz, J.; Radlein, D. & Majerski, P., 1996. Process for the production of anhydrosugars from lignin and cellulose containing biomass by pyrolysis. US5395455.

Sarotti, A.M.; Spanevello, R.A. & Suarez, A.G. (2007). An efficient microwave-assisted green transformation of cellulose into levoglucosenone. Advantages of the use of an experimental design approach. *Green Chem*, 9, 1137-1140.

Scholze, B. & Meier, D. (2001). Characterization of the water-insoluble fraction from pyrolysis oil (pyrolytic lignin). Part I. PY-GC/MS, FTIR, and functional groups. *J. Anal. Appl. Pyrol.*, 60, 41-54.

Scholze, B.; Hanser, C. & Meier, D. (2001). Characterization of the water-insoluble fraction from fast pyrolysis liquids (pyrolytic lignin) Part II. GPC, carbonyl goups, and C-13-NMR. *J. Anal. Appl. Pyrol.*, 58, 387-400.

Shafizadeh, F. & Lai, Y.Z. (1972). Thermal degradation of 1,6-anhydro-β-D-glucopyranose. *J. Org. Chem.*, 37, 278-284.

Shafizadeh, F. (1982). Introduction to pyrolysis of biomass. *J. Anal. Appl. Pyrol.*, 3, 283-305.

Shen, D.K. & Gu, S. (2009). The mechanism for thermal decomposition of cellulose and its main products. *Bioresour. Technol.*, 100, 6496-6504.

Shen, D.K.; Gu, S. & Bridgwater, A.V. (2010). Study on the pyrolytic behaviour of xylan-based hemicellulose using TG-FTIR and Py-GC-FTIR. *J. Anal. Appl. Pyrol.*, 87, 199-206.

Shimada, N.; Kawamoto, H. & Saka, S. (2008). Different action of alkali/alkaline earth metal chlorides on cellulose pyrolysis. *J. Anal. Appl. Pyrol.*, 81, 80-87.

Stradal, J.A. & Underwood, G.L., 1995. Process for producing hydroxyacetaldehyde. US5393542.

Torri, C.; Lesci, I.G. & Fabbri, D., (2009a). Analytical study on the pyrolytic behaviour of cellulose in the presence of MCM-41 mesoporous materials. *J. Anal. Appl. Pyrol.*, 85, 192-196.

Torri, C.; Lesci, I.G. & Fabbri, D. (2009b). Analytical study on the production of a hydroxylactone from catalytic pyrolysis of carbohydrates with nanopowder aluminium titanate. *J. Anal. Appl. Pyrol.*, 84, 25-30.

Wan, Y.Q.; Chen, P.; Zhang, B.; Yang, C.Y.; Liu, Y.H.; Lin, X.Y. & Ruan, R. (2009). Microwave-assisted pyrolysis of biomass: Catalysts to improve product selectivity. *J. Anal. Appl. Pyrol.*, 86, 161-167.

Wang, C.; Hao, Q.L.; Lu, D.Q.; Ja, Q.Z.; Li, G.J. & Xu, B. (2008). Production of light aromatic hydrocarbons from biomass by catalytic pyrolysis. *Chin. J. Catal.*, 29, 907-912.

Williams, P.T. & Nugranad, N. (2000). Comparison of products from the pyrolysis and catalytic pyrolysis of rice husks. *Energy*, 25, 493-513.

Zeitsch, K.J. (2000). The chemistry and technology of furfural and its many by-products. Elsevier publication.

Zhang, H.Y.; Xiao, R.; Wang, D.H.; Zhong, Z.P.; Song, M.; Pan, Q.W. & He, G.Y. (2009). Catalytic fast pyrolysis of biomass in a fluidized bed with fresh and spent fluidized catalytic cracking (FCC) catalysts. *Energ Fuel*, 23, 6199-6206.

# 9

# Supercritical Water Gasification of Biomass and Organic Wastes

Liejin Guo, Changqing Cao and Youjun Lu
*State Key Laboratory of Multiphase Flow in Power Engineering (SKLMF)*
*Xi'an Jiaotong University, Xi'an 710049,*
*PR China*

## 1. Introduction

With the exhaustion of fossil fuels and the severe pollution of the environment, it's urgent to find substitute renewable energy resources. Biomass is a kind of widespread and abundant renewable resource with lower sulfur content than fossil fuels, so the pollution to environment from biomass usage is much slighter than that from fossil fuels usage. Besides, zero emission of $CO_2$ can be realized in the cycle of growing and energy conversion of biomass.

Supercritical water gasification (SCWG) is an innovation biomass conversion process which takes advantage of the special properties of supercritical water (temperature above 374°C and pressure above 22.1MPa) to transform biomass into hydrogen-rich gaseous products. Comparing with conventional gasification and pyrolysis in normal pressure, SCWG has many advantages:

- Most organic materials of biomass can be dissolved in supercritical water for the relatively high dielectric constant of supercritical water (SCW), thus SCWG of biomass is homogeneous reaction, which can decrease the mass transfer resistance between phases.
- High pressure of the gaseous product make it easy for transportation, usage, carbon capture and further purification of the product gas through steam reforming or PSA (pressure swing adsorption).
- Higher energy efficiency can be achieved in SCWG of biomass especially for high moisture content biomass as the drying process is not required in SCWG.
- The reaction temperature is much lower than that in conventional gasification and pyrolysis. For example, the temperature of conventional steam gasification is always above 1000°C, while the complete gasification of glucose can be achieved at 650°C, 35.4MPa in supercritical water gasification.
- The gaseous product can be very clean. As almost no NOx and SOx were generated in supercritical water gasification, and the CO concentration is very low, especially with the catalyst to enhance the water-gas shift reaction.

Supercritical water gasification was firstly described by Modell in reforming of glucose and wood residues (Modell 1977; Modell 1980). In recent decades, much important progress has been made in SCWG technology by the researchers around the world. Elliott et al in PNNL (Pacific Northwest National Laboratory) have done a series of research on the reactions of kinds of biomass and organic wastes in high-pressure aqueous environments since 1980s.

They studied the effect of various catalysts on the reaction under the conditions around 350˚C, 20MPa and the methane-rich gaseous products was achieved in these conditions. Antal's group at the University of Hawaii proposed activated carbon as the catalyst in SCWG and realized complete gasification of biomass in 650°C, 34.5MPa. The researchers in FK(Forschungszentrum Carlsruhe) of Germany have done much research on SCWG since 2000, and found that the addition of alkali salt can not only increase the reaction rate and hydrogen yield but also can inhibit the generation of tar and char. In 2005, an excellent review about the status of biomass supercritical water gasification(Matsumura et al, 2005) was reported by several researchers. Since 1997, Guo's group in SKLMF of Xi'an Jiaotong University has done a series of research on supercritical water gasification of biomass and much progress on SCWG of biomass has been made, including the gasification of various feedstocks, the system development and the exploration of the influence rule of the various parameters.

In this chapter, the progress have been made on SCWG in recent decades will be reviewed. Firstly, the different biomass feedstocks used in supercritical water will be introduced, which include the model compound, real biomass and organic wastes. Secondly, the engineering problems in the supercritical water gasification process will be discussed and the solutions to these problems will also be presented. Thirdly, the influence factors on supercritical water gasification will be discussed and the relevant investigation results will be also introduced. Then a brief review of the catalyst used in supercritical water gasification will be presented. However, there are still some challenges to the supercritical water gasification technology, which will be discussed in the end.

## 2. The feedstock materials

### 2.1 Model compound

Using a model compound can help us to understand the basic chemical pathways occurred in supercritical water gasification. The main components of biomass are cellulose, hemicellulose and lignin. Cellulose and hemicellulose are carbohydrate while lignin includes aromatic rings.

Glucose is always used as a model compound for cellulose for these reasons: Firstly, as the main component of biomass, cellulose consists of linearly linked glucose units attached to each other; Secondly, glucose is the primary product of cellulose hydrolysis around the critical point of water (Sasaki et al, 1998) and the gasification of cellulose and glucose have the similar gasification products. In addition, glucose is easy to transport in the high pressure for its water-soluble matter property.

### 2.2 Real biomass

Besides the three main components (cellulose, hemicellulose and lignin), real biomass also contains other substances, such as alkali salt, sulphur and proteins. Supercritical water gasification of different real biomass may have different gasification results because of the different components, which will be described below.

In our previous study, the real biomass including wood sawdust, rice straw, rice shell, wheat stalk, peanut shell, corn stalk, corn cob and sorghum stalk were gasified in supercritical water and the results were shown in fig.1 (Lu et al, 2006). It can be found that the gasification results are different for various biomass feedstocks. Wheat stalk, corn cob and sorghum stalk are easier to gasify than the other biomass. The unconverted TOC (Total organic carbon) in liquid effluent was high, which indicated that a portion of biomass was converted to liquid products instead of gas products. This difference may be associated with the different amount of the components mentioned above.

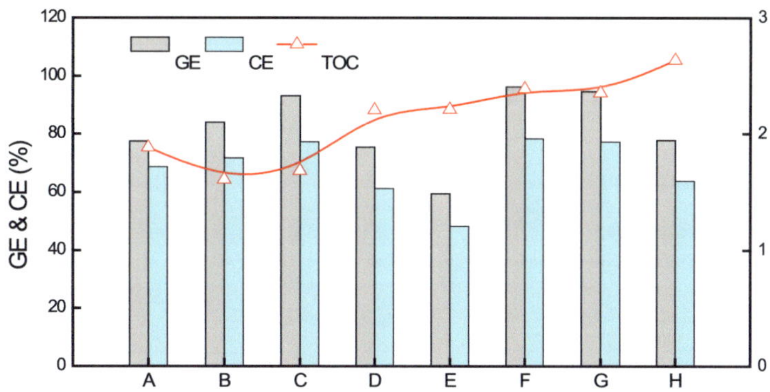


Fig. 1. Comparison of GE, CE and TOC for gasification of various biomass feedstocks in SCW (Temperature, 650°C; Residence time, 27s; Pressure, 25MPa; Feedstock, 2 wt% biomass +2wt% CMC). A, rice straw; B, rice shell; C, wheat stalk; D, peanut shell; E, corn stalk; F, corn cob; G, sorghum stalk; H, wood sawdust.(Lu et al, 2006)

### 2.3 Organic wastes

Supercritical water gasification of organic wastes can realize the hydrogen production and decontamination simultaneously. Besides, the homogeneous solution of wastewater makes it easy to pump to the high pressure reactor without pretreatment. In recent years, supercritical water gasification of various kinds of organic wastes was investigated by the researchers. Xu and Antal (1998) gasified 7.69wt% digested sewage sludge in supercritical water by mixing with corn starch gel to form a viscous paste. They found that the digested sewage sludge was gasified to a gas composed of $H_2$, $CO_2$, $CH_4$, and a trace of CO. A carbon gasification ratio as high as 98% was achieved in their studies. The aqueous effluent from the reactor had a low TOC value, a neutral pH, and no color. But the plugging problem in the reactor in supercritical water gasification of sewage sludge occurred due to the high ash content of the material. The gasification of many kinds of organic wastes in hot-compressed water (around 350°C, 20MPa) was investigated in PNNL (Elliott et al, 1994; Elliott et al, 2006) and methane was produced as the main product. The gasification of waste plastic in supercritical water also attracted much attention. Supercritical water gasification of waste plastics and the model compounds (such as polyethylene) were investigated by many researchers (Watanabe et al, 1998; Watanabe et al, 2001; Shibasaki et al, 2004; Su et al, 2004; Takeshita et al, 2004; Su et al, 2007).

The anaerobic organic wastewater from wheat straw includes acids (acetic acid, butyric acid) and ethanol etc. The gasification of anaerobic organic wastewater in supercritical water was investigated with a continuous operation tube flow reactor system (Ji 2006). The details of the reactor system were described in our previous report (Guo et al, 2007). The effect of temperature on SCWG of anaerobic organic wastewater was shown in fig. 2. It is showed that the main product gas are $H_2$, $CO_2$ and $CH_4$, with trace amount of CO and $C_2$. The gasification efficiency and carbon gasification efficiency were relatively high, for example,

100% gasification efficiency was achieved at the temperature of 775°C. The gasification efficiency, carbon gasification efficiency and the total gas yield increased with the temperature increased from 700°C to 775°C. The results show that the combination of the anaerobic hydrogen production of raw biomass with supercritical water gasification process can not only get hydrogen but also reduce the pollution.

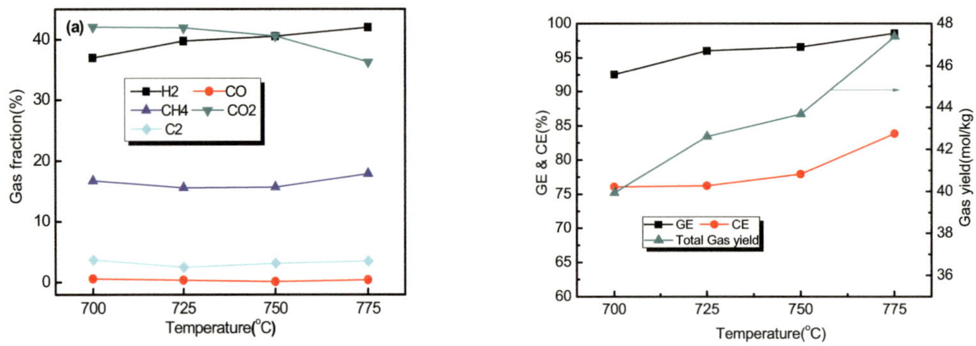


Fig. 2. The effect of temperature on SCWG of 6wt% anaerobic organic wastewater at 25MPa: (a) Gas fractions; (b)GE, CE & total gas yield.

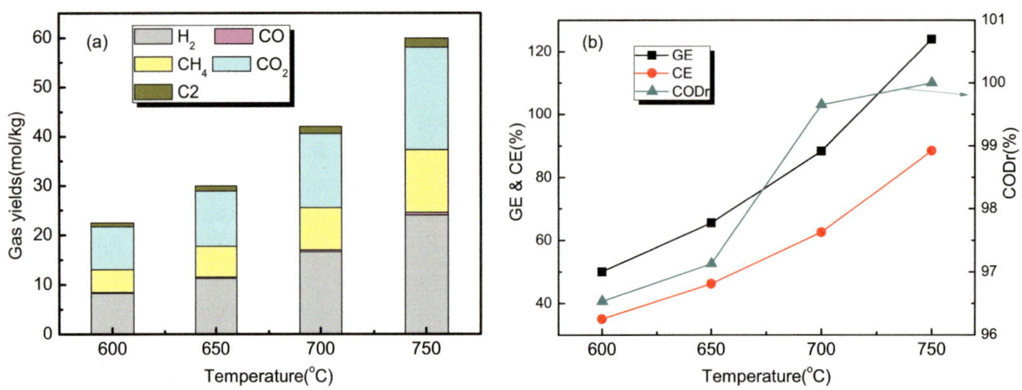


Fig. 3. The effect of temperature on SCWG of 7.8wt% black liquor at the pressure of 22.5MPa: (a)gas yields; (b)GE, CE and CODr

Black liquor is a kind of byproduct in pulping process and contains about 90% COD concentration of the pulping wastes. It mainly contains lignin derivatives which are hard to degrade and high content of alkali wastes. It is a wastewater with dark color, odor and high alkalinity. The black liquor contained 7.8wt% solid material was gasified in supercritical water with the same continuous operation tube flow reactor at the pressure of 22.5MPa. The gasification results with different temperature were shown in fig. 3. With the increasing of the temperature form 600°C to 750°C, the gas yield doubled and the COD removal efficiency, gasification efficiency and carbon gasification efficiency also increased significantly. The maximal gasification efficiency (123%) and carbon gasification efficiency (88%) were achieved at the temperature of 750°C. The COD content was fully removed, which means that the complete decontamination of black liquor can be achieved at the temperature of 750°C.

## 3. Engineering problems in supercritical water gasification systems

As an innovation biomass processing technology, supercritical water gasification has many new engineering problems, especially because the reaction temperature and pressure are relatively high. During decades of development on supercritical water gasification, much progress has been made in the engineering of supercritical water gasification process. Fig. 4 displays a typical continuous SCWG experimental system developed in SKLMF. Taking this system as example, some engineering problems will be discussed as below.

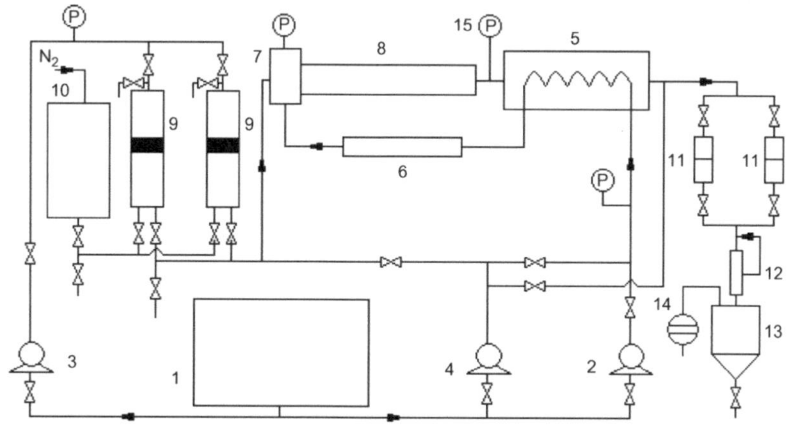


Fig. 4. Schematic diagram of bench-scale continuous SCWG apparatus: (1) Water tank; (2) Preheated water pump; (3) Feed pump; (4) Wash pump; (5) Cooler and exchanger; (6) Pre-heater; (7) mixer ; (8) Reactor; (9) Feeder; (10) Feed tank; (11) Filter; (12) Back-pressure regulator; (13) Gas-liquid separation; (14) Gas meter; (15 ) Pressure transducer.(Li et al, 2010)

### 3.1 Heating rate

The feedstock heating rate is reported to have significant influence on supercritical water gasification. Xu et al found that improving heating rate of the feedstock delays deactivation of coconut shell activated carbon catalyst, likely for the inhibition of tar formation (Xu et al, 1996). Kruse et al found that coke/char was generated with the slower feedstock heating rate (Kruse et al, 2003). They also gave the explanation for the formation of coke at low heating rates: When the biomass/water mixture spends enough time at sub-critical temperatures, furfurals or other unsaturated compounds are formed in significant yields. These compounds may polymerize as the temperature increases (Kruse 2008). Matsumura's group investigated SCWG at different feedstock heating rate and found that a heating rate of several hundreds of degrees Kelvin per minute should be desirable for the inhibition of coke formation (Matsumura et al, 2005).

In this system (fig.4), we realized the fast heating by mixing the feedstock with the water preheated by the heat exchanger (No. 5 in fig.4) and pre-heater(No. 6 in fig.4). Thus the coke formation can be inhibited as the feedstock can be heated to supercritical condition at the inlet of the reactor.

### 3.2 Energy recovery

High temperature reactor effluent owns a great quantity of enthalpy, so the recovery of the energy is necessary for the SCWG process. It is reported that the heat needed for supercritical water gasification can not be supplied by the heating value of the feedstock. From the heat balance considerations, it is clear that heat exchange between the reactor effluent and the reactor feed is essential for the economics of the process (Matsumura et al, 2005).

The effects of the heat transfer efficiency of the heat exchanger between the reactor effluent and the reactor feed (No.5 in fig. 4) on total energy and exergy efficiencies were analyzed, and the results were shown in fig. 5. As expected, total energy and exergy efficiencies of the biomass gasification increased with the increase of heat transfer efficiency in the heat exchanger. The increasing tendency is even more obvious with higher heat transfer efficiency of the heat exchanger.

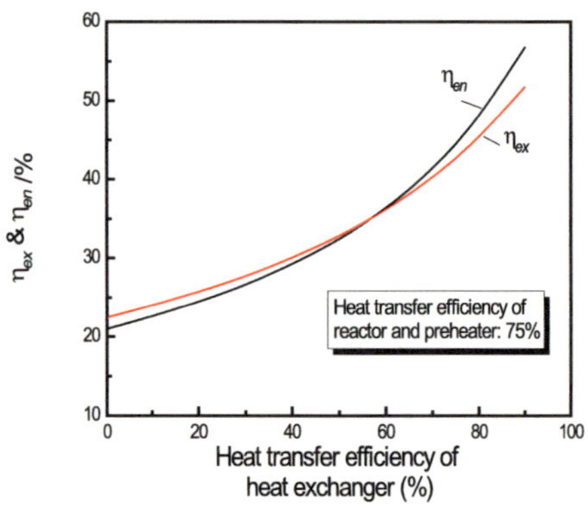


Fig. 5. The energy and exergy conversion efficiency of SCWG system with different heat transfer efficiency of heat exchanger. (Lu et al, 2007)

### 3.3 Plugging problem

Plugging is always a problem in biomass supercritical water gasification in tubular reactors (Antal et al, 2000). There are two reasons for plugging problem in SCWG system: The first one is the coke generated for the incomplete gasification of biomass; Secondly, as mentioned above, the solubility of inorganic salts is very low in the supercritical water and most of the salts were precipitated in the water. The plugging problem can cause the shutdown of the system which presents a severe problem for the energy conversion process (Matsumura and Minowa, 2004).

To handle with the plugging problem in SCWG of biomass, especially for the gasification of high concentration biomass, a novel SCW fluidized bed system for biomass gasification was developed in SKLMF. The details of this system were described in our previous paper (Lu et al, 2008). The innovation of the system is the application of fluidized bed reactor and the fluidizing agent is $SiO_2$ beads with diameters in the range of 0.1–0.15mm. In this system, high concentration model biomass (30wt% glucose) and real biomass (18wt% corn cob) feedstock were gasified successfully without plugging.

### 3.4 Continuous feeding

The delivery of biomass to the supercritical water gasification reactor is a technological challenge, for the precipitation will occurred in the continuous feeding for some biomass which can not dissolve in the water. The solution to this problem is to increase the viscosity of the slurry, Matsumura et al successfully applied a pre-treatment method in hot compressed water (150°C, 30min) for feeding cabbage(Matsumura et al, 2005). They found the softening of the hard structure of biomass as observed in making soup in the kitchen. And this 'softening' effect increased with the increase of the temperature and pressure. Antal's group present a solution which suspended the biomass in a starch gel and successfully delivered the sawdust to the reactor via a 'cement pump'(Antal et al, 2000). In SKLMF, the sodium carboxymethylcellulose (CMC) was chosen as an additive to mix with the solid biomass feedstock and water, and realized the continuous pumping of the sawdust (Hao et al, 2003).

### 3.5 The separation of products

The main product gas compositions of biomass SCWG are $H_2$, $CH_4$, CO and $CO_2$. To remove carbon dioxide to raise the heating value of the product gas, separation of gas and liquid before depressurization is effective. Based on the higher solubility of carbon dioxide than hydrogen in water, utilization of high-pressure separator for hydrogen and carbon dioxide was proposed. Above 90mol% hydrogen purity gas phase was achieved with the addition of excess water to the product gas mixture under high pressure (Matsumura et al, 1997a).

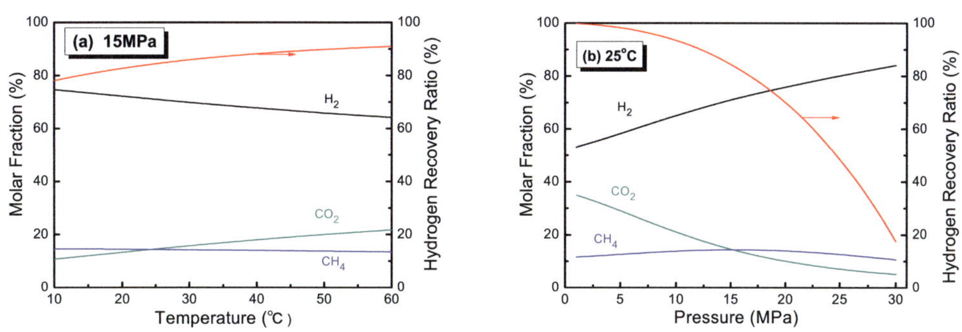


Fig. 6. The gas fractions and the hydrogen recovery ratio in the gas phase of the high pressure separator with different: (a)pressure; (b)temperature(Lu et al, 2007).

The gas-liquid equilibrium of the products in the high pressure separator was analyzed with the assumption that the chemical equilibrium is reached in the reactor at 600°C, 25MPa. The effects of the operation temperature and pressure of the high pressure separator on the gas fractions and hydrogen recovery ratio in the gas phase were studied. The results shown in fig.6 revealed that as operation temperature increases, the molar fraction of $H_2$ in gas phase decreased while the $CH_4$, $CO_2$ fractions and hydrogen recovery ratio increased. Purity of $H_2$ in the gas phase is 86.24% at 283K and 75.7% at 333 K, respectively. As a result, proper operation temperature of the high pressure separator should be selected to consider both $H_2$ purity and hydrogen recovery ratio. Fig. 6(b) shows that the molar fraction of hydrogen in the gas phase increases from 65.56% to 92.41% and the molar fraction of $CO_2$ decreases sharply from 33.11% to 6.12% with the pressure increasing from 0.1MPa to 30MPa. Hydrogen recovery ratio decreased and the molar fraction of $CH_4$ increased a little with the increasing pressure. The results suggested that the increase of pressure in the separator favors the purity of $H_2$ in the gas phase but decreases the hydrogen recovery ratio, so appropriate operation pressure of high pressure separator must be selected. The predicted results show that $H_2$ of 82.45% and recovery ratio of 88.15% are obtained at 15MPa, 298 K.

$H_2$ and $CH_4$ in the liquid phase can be separated in a low pressure separator and combust with oxygen to produce heat, which can be recovered for the gasification system to reduce external heat input.

## 4. The influence factors on supercritical water gasification

### 4.1 The influence of main operating parameters

Based on minimizing Gibbs free energy principle, chemical equilibrium of sawdust SCWG was predicted by thermodynamic calculation code (Yan et al, 2006; Lu et al, 2007). In order to simplify the problem, nitrogen, sulfur and other impurities contained in the raw materials are assumed to be neglected, so wood sawdust is represented by a general formula of $CH_{1.35}O_{0.617}$. The predicted results show that the product gas includes mainly $H_2$, $CH_4$, CO and $CO_2$. The influence of the main operating parameters in SCWG was predicted and shown in fig.7-10.

*(a) Influence of temperature*

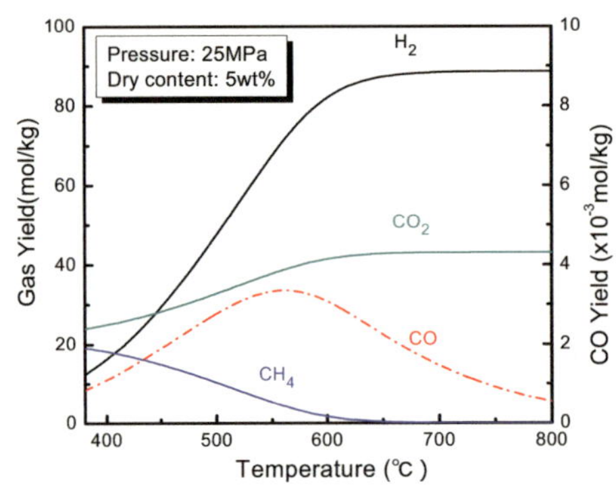


Fig. 7. Equilibrium gas yields of SCWG of 5wt% sawdust with change of temperature.

Fig. 7 shows the equilibrium gas yields of sawdust gasification as a function of reaction temperature at 25 MPa. At the chemical equilibrium state, the yields of $H_2$ and $CO_2$ increase with the increasing temperature, but the yield of $CH_4$ decreases sharply. The equilibrium CO yield is very small, and it is about $10^{-3}$ mol/kg. As temperature increases from 400 to 800°C, the CO yield firstly increases and then drops down. The maximum CO yield is reached at about 550°C. Hydrogen yield increases at a low speed at rather higher temperature. When the reaction temperature is above 650°C, biomass gasification goes to completion and the equilibrium gas product consists of $H_2$ and $CO_2$ in a molar ratio equal to $(2 - y + x/2)$.( x and y are the elemental molar ratios of H/C and O/C in biomass, respectively). The maximal equilibrium $H_2$ yield of 88.623 mol/kg dry biomass is obtained. From the viewpoint of thermodynamics, higher temperature is essential to hydrogen production.

As shown above, temperature has significant effect on biomass gasification in SCW and this conclusion was confirmed by the experimental results. Xu et al (1996) reported that 1M glucose was gasified completely at 600°C, and a thin layer of dark brown oil-like tar was observed at the temperature below 580°C. For the gasification of high concentration feedstocks, the temperature of 650–800°C is needed (Antal et al, 2000). Further more, the higher temperature drove the methane steam-reforming reaction to increase hydrogen yield (Sealock et al, 1993).

*(b) Influence of pressure*

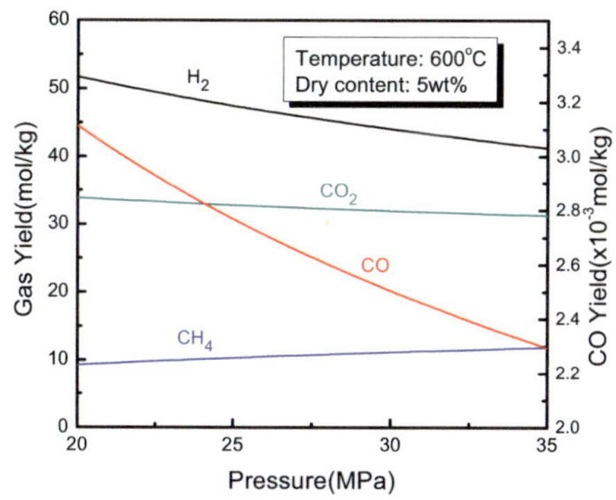


Fig. 8. Equilibrium gas yields of SCWG of wood sawdust with change of pressure.

Pressure shows a complex effect on biomass gasification in SCW. The properties of water, such as density, static dielectric constant and ion product increase with pressure. As a result, the ion reaction rate increases and free-radical reaction is restrained with an increase of pressure. Hydrolysis reaction plays a significant role in SCWG of biomass, which requires the presence of $H^+$ or $OH^-$. With increasing pressure, the ion product increases, therefore the hydrolysis rate also increase. Besides, high pressure favors water–gas shift reaction, but reduce decomposition reaction rate.

Fig. 8 shows the effect of pressure on equilibrium gas yield at 600°C with 5 wt% biomass content. It reveals that the pressure has no significant effect on equilibrium gas yield. The similar experimental results were achieved when we gasified the 2wt% wood sawdust in supercritical water at the temperature of 650°C, in the pressure range of 18-30MPa (Lu et al, 2006). It is found that the hydrogen yield, GE and CE is not monotonic functions of pressure when the pressure is near the critical pressure, but the hydrogen yield, GE and CE increase a little as pressure increases from 25 to 30 MPa. Demirbas (2004) also found that hydrogen yield increased as pressure increased from 23 to 48 MPa in SCWG of fruit shell and it is considered that the increase of the pressure increased the mass transfer and the solvent diffusion rates of the water. Thus the gasification efficiency of supercritical water gasification increased with the pressure.

*(c) Influence of feedstock concentration*

Fig. 9 displays the effect of feedstock concentration on equilibrium gas yield at 600°C and 25 MPa. The product gas mainly consists of $H_2$ and $CO_2$ when biomass feedstock with low concentration is gasified, but the $CH_4$ yield is very high when the high concentration feedstock is gasified.

The similar results were achieved in the SCWG of wood sawdust in a batch reactor (Lu et al, 2006). The gasification results showed that the dry biomass content has significant effect on biomass gasification and the high concentration feedstock is more difficult to gasify. With higher feedstock concentration, the gasification efficiency and $H_2$ yield decreased, while the

CO yield increased. But the gasification of high concentration feedstock is necessary to achieve a thermal efficiency high enough to establish an economic process. For high efficiently gasification of high concentration feedstock in supercritical water, the high temperature, high heating rate and catalyst are required (Antal et al, 2000).

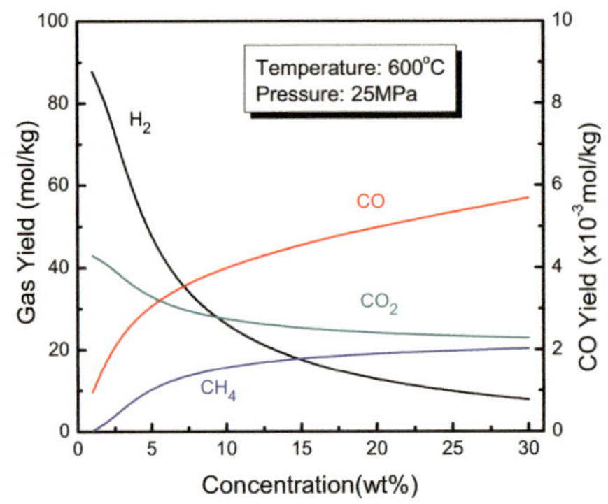


Fig. 9. Equilibrium gas yields of SCWG of wood sawdust with change of concentration.

*(d) Influence of the oxidant*

The oxidant is expected to decompose the complex compound of the reactant or the intermediate products in supercritical water, such as the phenols which is considered to be 'the last hurdle' to get over to complete gasification of biomass (Kruse et al, 2003).Thus the formation of tar and char can be decreased. In addition, the in-situ heat generated from the oxidation reaction can heat the feedstock rapidly, which favors the process of gasification (Watanabe et al, 2003; Matsumura et al, 2005). The effect of oxidant addition on equilibrium gas yield was predicted and the results were shown in fig.10. It revealed that with the increase of the oxidant addition, the yields of $H_2$, CO and $CH_4$ decreased and the yield of $CO_2$ increased. The addition of oxidant can enhance the efficiency of biomass SCWG and provide the heat for the reactions in SCW, but decreased the hydrogen yield.

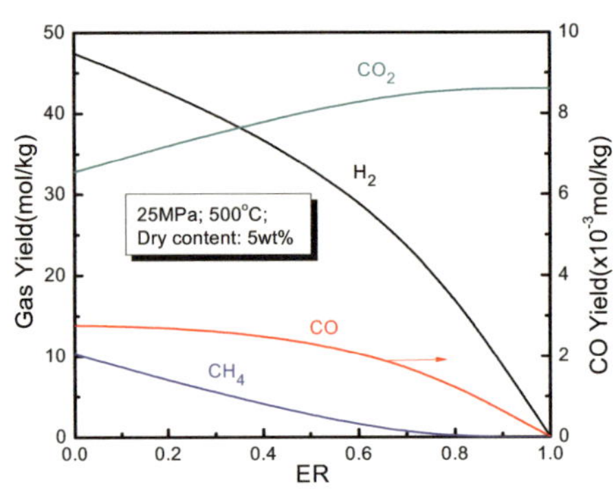


Fig. 10. Equilibrium gas yields of SCWG of wood sawdust with change of oxidant addition.

The influence of the oxidant addition on SCWG of model compounds (glucose, lignin) was investigated in a fluidized bed system (Jin et al, 2010). The results showed that the oxidant can improve the gasification efficiency and an appropriate addition of oxidant can improve the yield of hydrogen in certain reaction condition. When ER equaled 0.4, the gasification efficiency of lignin was 3.1 times of that without oxidant. When ER equaled 0.1, the yield of hydrogen from lignin increased by 25.8% compared with that without oxidant. But when the oxidant addition increased to a certain level, the cold gas efficiency decreased for the consumption of the combustible gas in the oxidation reaction. So there is an optimum oxidant addition amount in SCWG.

*(e) Influence of reaction time*

From the viewpoint of thermodynamics, biomass can be gasified completely in SCW with a product formation of $H_2$ and $CO_2$, but adequate reaction time was required to complete the gasification process. Antal et al (1994) gasified 0.1 M glucose at 34.5 MPa, 600°C with various residence times. They found that glucose can be gasified quickly and the complete gasification was achieved in only 28 s residence time. Lee et al (2002) reported the yields of all the gases remained almost constant at 700°C, being independent of the residence time except for the shortest residence time of 10.4 s when the 0.6 M glucose was gasified at 28 MPa.

### 4.2 Influence of biomass properties

*(a) Influence of the main component*

As mentioned above, cellulose, hemicellulose and lignin are the main components of the biomass and they have different structures. So the different components may have different effect on SCWG. Yoshida et al investigated the interaction of cellulose, xylan(model compound for hemicellulose) and lignin by mixing them in different ratios in SCWG (Yoshida and Matsumura 2001). They found that the hydrogen yield by SCWG of the cellulose and hemicellulose are higher than that of lignin, there was no apparent interaction between the hydrogen production from cellulose and hemicellulose in SCWG. While with the mixing with lignin, the hydrogen production from SCWG of cellulose and hemicellulose was suppressed. In the following article (Yoshida et al, 2003), they showed that this effect depended on the species of lignin and the interaction between each component was also observed in the real biomass feedstocks (sawdust and rice straw). This result reveals the gasification of various biomass in SCW may have different results for their different components.

*(b) Influence of the protein content*

The proteins are contained in some biomass, such as the food industry residues or sewage sludge. Kruse et al (2005) studied the effect of proteins on hydrothermal gasification of biomass by comparison of the gasification results of two biomass feedstocks: One biomass feedstock originated from plant material (phyto mass) nearly contains no proteins and the other contains protein (zoo mass). They found that gas yield from SCWG of protein containing biomass (zoo mass) was much lower than that of phyto mass without proteins. To understand these findings, they conducted the experiments with the alanine as the model compound of protein (Kruse et al, 2007). The results showed that with the presence of alanine, the gas yield of glucose was decreased and the gas composition and the concentration of key compounds are slightly changed. They inferred that the nitrogen containing cyclic organic compounds was produced from the Maillard reaction between glucose and amino acids or their consecutive products. And these compounds were believed to be strong free radical scavengers and inhibit free radical chain reactions that are highly relevant for gas formation. In addition, experiments with real biomass in a batch reactor were reported to verify the assumption that Maillard products reduced free radical reactions. As an example, the addition of urea to phyto mass leads to a decrease of the gas yield to a value similar to that found for zoo mass. Dileo et al. (2008) examined the gasification of glycine as the model compound of protein in supercritical water. They found that glycine was much more resistant to be gasified than phenol. Large amounts (20%-90%) of the initial carbon remained in the aqueous phase even after 1 h for both homogeneous and Ni-catalyzed reactions.

*(c) Influence of inorganic elements*

The $K_2CO_3$ content of real biomass is always slightly higher than 0.5wt% (Sinag et al, 2003). The alkali is advantageous for SCWG as a catalyst. The addition of alkali salts can enhance the water-gas shift reaction in supercritical water gasification which resulted in higher $H_2$ fraction and lower CO fraction in the product gas. Also the alkali salts can also lead to more liquid product and less coke/char formation. The detail of the alkali catalysis effect will be described in section 5.

Sulfur also exists in some waste biomass and it has an influence on supercritical water gasification. Elliott et al claimed that the presence of sulfur lowered the activities of ruthenium catalysts in subcritical water at 623 K(Elliott et al, 2004). Osada et al studied the effect of sulfur on SCWG of lignin at 673K with the catalysis of supported ruthenium (Osada et al, 2007a). They found that the gas yield decreased with the increase of the sulfur added. The carbon dioxide fraction in the presence of sulfur was lager than that without sulfur, and the methane fraction was lower. From X-ray photoelectron spectroscopy characterization of catalysts used for gasification, the sulfur species which poisoned the ruthenium sites were found to be ruthenium sulfide, ruthenium sulfite, and ruthenium sulfate. In the further study about the effect on SCWG of lignin with $Ru/TiO_2$, they come to a conclusion that sulfur poisoned the active sites for carbon-carbon bond breaking and methanation reaction; on the other hand, it did not hinder the sites for the gasification of formaldehyde and the water-gas shift reaction (Osada et al, 2007b). Therefore, the desulfurization from biomass, especially the biomass waste, is essential for the development of the supercritical water catalytic gasification.

*(d) Influence of biomass particle size*

Biomass was pretreated with mechanical grinding before gasification. Biomass with different particle sizes were gasified in supercritical water in a batch reactor and the results showed that higher hydrogen yield is obtained with gasification of smaller particle size (Lu et al, 2006). We inferred that with larger particle size, the diffusion resistance may be larger and decreased the gasification efficiency. More energy will be consumed to achieve the smaller particle size for the mechanical grinding, so an optimal particle size should be found with considering economy and feasibility of the process.

## 5. Review of SCWG catalyst

To improve the economical efficiency of SCWG, the improvement of gasification efficiency as well as lowering the operating temperature should be considered. For this purpose, catalyst is one possible solution. Various catalysts were used for biomass thermal chemical gasification and a review of literature on catalysts for biomass gasification was published in 2001 (Sutton et al, 2001). However, the catalyst for SCWG should be different from the conventional gasification because of the particular operating conditions, such as the high pressure values, the purpose(hydrogen production instead of syngas) and the specificities of the supercritical medium (Calzavara et al, 2005). Generally, four types of catalysts were used for SCWG in the literature: alkali, activated carbon, metal and metal-oxide.

### 5.1 Alkali

The addition of alkali, such as NaOH, KOH, $Na_2CO_3$, $K_2CO_3$ and $Ca(OH)_2$ has significant influence on SCWG of biomass. Watanabe et al (2003) studied catalytic effects of NaOH in partial oxidative gasification of n-hexadecane and lignin in supercritical water (40MPa, 400°C). The results showed that the $H_2$ yield with NaOH was almost 4 times higher than that without catalyst. Kruse et al (2000) conducted SCWG of different organic compounds in the presence of KOH. They found that the addition of KOH decreased the CO fraction and increased the fractions of hydrogen and carbon dioxide by accelerating of water-gas shift reaction. The similar results were achieved by Sinag et al(Sinag et al, 2003; Sinag et al, 2004) when they gasified glucose in SCW with 0.5wt% $K_2CO_3$. They also regarded that the formation of the formate salt was the reaction mechanism of the acceleration of the water-gas shift reaction by alkali salts in SCWG. The alkali is also well-known as the catalyst for the oil production from biomass, where their important role is to inhibit the char formation from the oil (Minowa et al, 1998). Thus, alkali has a positive effect to produce gaseous product such as $H_2$. Furthermore, the addition of the $Ca(OH)_2$ can not only catalysis the SCWG of biomass as described above, but it can also adsorb $CO_2$ to decrease the $CO_2$ fraction in the product gas(Lin et al, 2001; Lin et al, 2002; Lin et al, 2003; Lin et al, 2005). The high hydrogen purity gases were produce from this process.

### 5.2 Activated carbon

Xu et al (1996) used carbon-based catalysts, such as coal activated carbon, coconut shell activated carbon, macadamia shell charcoal and spruce wood charcoal, for organic compounds gasification in SCW. Complete conversion of glucose was achieved at 600°C, 34.5MPa. Subsequently, Antal and Xu (1998) and Antal et al (2000) gasified the high concentration biomass feedstocks completely above 650°C with carbon-based catalyst in SCW. The produced gases were mainly composed of hydrogen and carbon dioxide and the extraordinary hydrogen yield could be more than 100 g/kg of dry biomass. The carbon is thought to react with supercritical water. However, the rate of the supercritical water gasification of activated carbon was found to be very slow under typical SCWG conditions (Matsumura et al, 1997b). For the notable catalysis effect on SCWG and the stability of the carbon in SCW, activated carbon is used widely as the catalyst and the catalyst support. The catalysis effect of Ru/C and Pb/C on gasification of cellulose and sawdust in SCW was examined in our laboratory and it was found that these catalysts have remarkable effect on SCWG. 10wt% cellulose or sawdust with CMC can be gasified near completely with Ru/C and 2-4g hydrogen yield and 11-15g potential hydrogen yield per 100g feedstock were produced at the condition of 500°C, 27MPa and 20min residence time in an autoclave reactor (Hao et al, 2005).

### 5.3 Metal catalyst

Metal is widely used as catalyst in biomass conventional gasification and supercritical water gasification. Elliott et al (Elliott et al, 1993; Elliott and Sealock 1996) demonstrated that Ru, Rh and Ni had significant activity for the conversion of *p*-cresol and Pt, Pd and Cu was reported to have less activity. Sato et al. (2003) gasified alkylphenols as lignin model compound in the presence of supported noble metal catalysts in SCW at 40°C. The activity of the catalyst was in the order of Ru/a-alumina> Ru/carbon, Rh/carbon > Pt/a-alumina, Pd/carbon and Pd/a-alumina. Usui et al (2000) presented Pd/$Al_2O_3$ had the highest catalytic activity for cellulose gasification among the supported Ni, Pd or Pt. Nickel is cheaper than noble metals, so it is more suitable for large-scale hydrogen production by biomass gasification. Elliott et al (1993) tested several forms of nickel catalysts at 350°C and 17–23 MPa using a batch reactor, and the $CH_4$-rich gas was produced. Minowa and co-workers (Minowa & Ogi, 1998; Minowa et al, 1998; Minowa and Inoue, 1999) studied the effect of a reduced nickel catalyst on cellulose decomposition in hot-compressed water. They found that the nickel catalyst can accelerate the steam reforming of aqueous products and the methanation reaction.

### 5.4 Metal oxide

Although metal-oxide is not usually employed as a catalyst for biomass gasification, It was reported that (Watanabe et al, 2002) the hydrogen yield and the gasification efficiency of glucose and cellulose gasification in SCW with zirconia was almost twice as much as that without catalyst. The similar results were found in the partial oxidative gasification of lignin and n-C16 in SCW (Watanabe et al, 2003). Park and Tomiyasu (Park & Tomiyasu 2003) achieved nearly complete gasification of aromatic compounds in SCW with stoichiometrically insufficient amounts of $RuO_2$. We examined the catalytic effect of $CeO_2$ particles, nano-$CeO_2$, and nano-$(CeZr)_xO_2$ on SCWG of cellulose in our previous study (Hao et al, 2005) and found that these metal-oxide has limited catalytic effect on SCWG.

## 6. Challenges and prospect

As described above, much progress has been made in biomass supercritical water gasification, but there are still some problems to be resolved:

- Optimizing the process parameters especially in view of higher feed concentration necessary to achieve a thermal efficiency high enough to establish an economic process.
- The high pressure in SCWG process brings about challenge for the catalyst, such as the durable and life time of the catalyst. So developing long-life and cheap catalyst is important to increase economical efficiency of SCWG through improving the gasification efficiency and lowering the gasification temperature. On the other side, the recycling of the catalyst, especially the water soluble catalysts have also to be handled to decrease the cost of the process.
- Detailed kinetics should be developed based on the gasification mechanism and reaction path to give guidance to the design of supercritical water gasification system. So the detailed gasification mechanism have to be explored, especially the qualitative and quantitative analysis of the intermediate and end products.
- The corrosion is an inevitable problem for biomass supercritical water gasification as the reactor was exposed in severe conditions. Besides, the compositions of the biomass and intermediate products are complex. So it is important to find a construction

material which is resistant to corrosion or find a way to protect the reactor material from contacting with the reactant and products.

The energy conversion from biomass will be more urgent as the fossil fuel is running shorter nowadays. Though there are so many problems, supercritical water gasification is still a promising biomass conversion technology for its advantages over conventional gasification process. Especially for the organic wastes, supercritical water gasification can realize both the goals of energy recovery and decontamination simultaneously.

## 7. Nomenclature

GE: gasification efficiency, the mass of product gas/the mass of feedstock, %;
CE: carbon gasification efficiency, carbon in product gas/carbon in feedstock, %;
CODr: COD removal efficiency, 1-COD of aqueous residue/COD of feedstock, %;
ER: oxidant equivalent ratio, amount of oxidant added/the required amount for complete oxidation by stoichiometry calculation, %;

## 8. Acknowledgement

This work was financially supported by the National Natural Science Foundation of China (Contracted No. 50821064) and the National Basic Research Program of China (Contracted No. 2009CB220000). And we gratefully thanks to other colleagues in State Key Laboratory of Multiphase Flow in Power Engineering(SKLMF) for their contributions to this work. The authors also thank Dr. Jiarong Yin, Dr. Simao Guo and Dr. Zhiwei Guo for their valuable suggestions.

## 9. References

Antal, M. J., Allen S. G., Schulman D., Xu X. D. & Divilio R. J. (2000). Biomass gasification in supercritical water. *Ind. Eng. Chem. Res* 39(11): 4040-4053.

Antal, M. J., MANARUNGSON S. & MOK W. S. (1994). Hydrogen production by steam reforming glucose in supercritical water. *Adv Thermochem Biomass Convers* 3(2): 1367-1377.

Antal, M. J. & Xu X. (1998). Hydrogen production from high moisture content biomass in supercritical water. *Proceedings of the 1998 U.S.DOE Hydrogen Program Review,*, National Renewable Energy Laboratory, USA.

Calzavara, Y., Joussot-Dubien C., Boissonnet G. & Sarrade S. (2005). Evaluation of biomass gasification in supercritical water process for hydrogen production. *ENERGY CONVERS. MANAGE.* 46(4): 615-631.

Demirbas, A. (2004). Hydrogen-rich gas from fruit shells via supercritical water extraction. *Int. J. Hydrogen Energy* 29(12): 1237-1243.

Dileo, G. J., Neff M. E., Kim S. & Savage P. E. (2008). Supercritical water gasification of phenol and glycine as models for plant and protein biomass. *Energy & Fuels* 22(2): 871-877.

Elliott, D. C., Hart T. R. & Neuenschwander G. G. (2006). Chemical processing in high-pressure aqueous environments. 8. Improved catalysts for hydrothermal gasification. *Industrial & Engineering Chemistry Research* 45(11): 3776-3781.

Elliott, D. C., Neuenschwander G. G., Hart T. R., Butner R. S., Zacher A. H., Engelhard M. H., Young J. S. & McCready D. E. (2004). Chemical processing in high-pressure aqueous environments. 7. Process development for catalytic gasification of wet biomass feedstocks. *Ind. Eng. Chem. Res* 43(9): 1999-2004.

Elliott, D. C. & Sealock L. J. (1996). Chemical processing in high-pressure aqueous environments: Low-temperature catalytic gasification. *Chemical Engineering Research & Design* 74(A5): 563-566.

Elliott, D. C., Sealock L. J. & Baker E. G. (1993). Chemical-Processing in High-Pressure Aqueous Environments .2. Development of Catalysts for Gasification. *Ind. Eng. Chem. Res* 32(8): 1542-1548.

Elliott, D. C., Sealock L. J. & Baker E. G. (1994). Chemical-Processing in High-Pressure Aqueous Environments .3. Batch Reactor Process-Development Experiments for Organics Destruction. *Industrial & Engineering Chemistry Research* 33(3): 558-565.

Guo, L. J., Lu Y. J., Zhang X. M., Ji C. M., Guan Y. & Pei A. X. (2007). Hydrogen production by biomass gasification in supercritical water: A systematic experimental and analytical study. *Catalysis Today* 129(3-4): 275-286.

Hao, X. H., Guo L. J., Mao X., Zhang X. M. & Chen X. J. (2003). Hydrogen production from glucose used as a model compound of biomass gasified in supercritical water. *Int. J. Hydrogen Energy* 28(1): 55-64.

Hao, X. H., Guo L. J., Zhang X. M. & Guan Y. (2005). Hydrogen production from catalytic gasification of cellulose in supercritical water. *CHEM ENG J* 110(1-3): 57-65.

Ji, C. M. Experimental Study on Hydrogen Production by Gasification of Biomass and Organic Wastes under Supercritical Water in a Continuous Tube Flow Reactor [D]. Xi'an: Xi'an Jiaotong University, 2006.

Jin, H., Lu Y., Guo L., Cao C. & Zhang X. (2010). Hydrogen production by partial oxidative gasification of biomass and its model compounds in supercritical water. *Int. J. Hydrogen Energy* 35(7): 3001-3010.

Kruse, A. (2008). Supercritical water gasification. *Biofuel Bioprod Bior* 2(5): 415-437.

Kruse, A., Henningsen T., Sınağ A. & Pfeiffer J. (2003). Biomass Gasification in Supercritical Water: Influence of the Dry Matter Content and the Formation of Phenols. *Ind. Eng. Chem. Res* 42(16): 3711-3717.

Kruse, A., Krupka A., Schwarzkopf V., Gamard C. & Henningsen T. (2005). Influence of proteins on the hydrothermal gasification and liquefaction of biomass. 1. Comparison of different feedstocks. *Ind. Eng. Chem. Res* 44(9): 3013-3020.

Kruse, A., Maniam P. & Spieler F. (2007). Influence of proteins on the hydrothermal gasification and liquefaction of biomass. 2. Model compounds. *Ind. Eng. Chem. Res* 46(1): 87-96.

Kruse, A., Meier D., Rimbrecht P. & Schacht M. (2000). Gasification of pyrocatechol in supercritical water in the presence of potassium hydroxide. *Ind. Eng. Chem. Res* 39(12): 4842-4848.

Lee, I. G., Kim M. S. & Ihm S. K. (2002). Gasification of glucose in Supercritical water. *Ind. Eng. Chem. Res* 41(5): 1182-1188.

Li, Y., Guo L., Zhang X., Jin H. & Lu Y. Hydrogen production from coal gasification in supercritical water with a continuous flowing system. *Int J Hydrogen Energy* doi: DOI: 10.1016/j.ijhydene.2009.07.023.

Lin, S. Y., Harada M., Suzuki Y. & Hatano- H. (2003). Comparison of pyrolysis products between coal, coal/CaO, and coal/Ca(OH)(2) materials. *Energy & Fuels* 17(3): 602-607.

Lin, S. Y., Harada M., Suzuki Y. & Hatano H. (2005). CO2 separation during hydrocarbon gasification. *Energy* 30(11-12): 2186-2193.

Lin, S. Y., Suzuki Y., Hatano H. & Harada M. (2001). Hydrogen production from hydrocarbon by integration of water-carbon reaction and carbon dioxide removal (HyPr-RING method). *Energy & Fuels* 15(2): 339-343.

Lin, S. Y., Suzuki Y., Hatano H. & Harada M. (2002). Developing an innovative method, HyPr-RING, to produce hydrogen from hydrocarbons. *ENERGY CONVERS. MANAGE.* 43(9-12): 1283-1290.

Lu, Y., Guo L., Zhang X. & Yan Q. (2007). Thermodynamic modeling and analysis of biomass gasification for hydrogen production in supercritical water. *CHEM ENG J* 131(1-3): 233-244.

Lu, Y. J., Guo L. J., Ji C. M., Zhang X. M., Hao X. H. & Yan Q. H. (2006). Hydrogen production by biomass gasification in supercritical water: A parametric study. *Int. J. Hydrogen Energy* 31(7): 822-831.

Lu, Y. J., Jin H., Guo L. J., Zhang X. M., Cao C. Q. & Guo X. (2008). Hydrogen production by biomass gasification in supercritical water with a fluidized bed reactor. *Int. J. Hydrogen Energy* 33(21): 6066-6075.

Matsumura, Y. & Minowa T. (2004). Fundamental design of a continuous biomass gasification process using a supercritical water fluidized bed. *Int. J. Hydrogen Energy* 29(7): 701-707.

Matsumura, Y., Minowa T., Potic B., Kersten S. R. A., Prins W., van Swaaij W. P. M., van de Beld B., Elliott D. C., Neuenschwander G. G., Kruse A. & Antal M. J. (2005). Biomass gasification in near- and super-critical water: Status and prospects. *Biomass Bioenerg* 29(4): 269-292.

Matsumura, Y., Minowa T. & Xu X. (1997a). High-pressure carbon dioxide removal in supercritical water gasfication of biomass. *Dev Thermochem Biomass Conversion* 2: 864–877.

Matsumura, Y., Xu X. & Antal M. J. (1997b). Gasification characteristics of an activated carbon in supercritical water. *Carbon* 35(6): 819-824.

Minowa, T. & Inoue S. (1999). Hydrogen production from biomass by catalytic gasification in hot compressed water. *Renewable Energy* 16(1-4): 1114-1117.

Minowa, T. & Ogi T. (1998). Hydrogen production from cellulose using a reduced nickel catalyst. *Catalysis Today* 45(1-4): 411-416.

Minowa, T., Zhen F. & Ogi T. (1998). Cellulose decomposition in hot-compressed water with alkali or nickel catalyst. *Journal of Supercritical Fluids* 13(1-3): 253-259.

Modell, M. (1977). Reforming of Glucose and Wood at Critical Conditions of Water. *Mechanical Engineering* 99(10): 108-108.

Modell, M. (1980). Reforming of Organic-Substances in Supercritical Water. *J ELECTROCHEM SOC* 127(3): C139-C139.

Osada, M., Hiyoshi N., Sato O., Arai K. & Shirai M. (2007a). Effect of sulfur on catalytic gasification of lignin in supercritical water. *Energy & Fuels* 21(3): 1400-1405.

Osada, M., Hiyoshi N., Sato O., Arai K. & Shirai M. (2007b). Reaction pathway for catalytic gasification of lignin in presence of sulfur in supercritical water. *Energy & Fuels* 21(4): 1854-1858.

Park, K. & Tomiyasu H. (2003). Gasification reaction of organic compounds catalyzed by RuO2 in supercritical water. *Chem Commun (Camb)* 6: 694-695.

Sasaki, M., Kabyemela B., Malaluan R., Hirose S., Takeda N., Adschiri T. & Arai K. (1998). Cellulose hydrolysis in subcritical and supercritical water. *J Supercritical Fuilds* 13(1-3): 261-268.

Sato, T., Osada M., Watanabe M., Shirai M. & Arai K. (2003). Gasification of alkylphenols with supported noble metal catalysts in supercritical water. *Ind. Eng. Chem. Res* 42(19): 4277-4282.

Sealock, L. J., Elliott D. C., Baker E. G. & Butner R. S. (1993). Chemical-Processing in High-Pressure Aqueous Environments .1. Historical-Perspective and Continuing Developments. *Ind. Eng. Chem. Res* 32(8): 1535-1541.

Shibasaki, Y., Kamimori T., Kadokawa J., Hatano B. & Tagaya H. (2004). Decomposition reactions of plastic model compounds in sub- and supercritical water. *Polymer Degradation and Stability* 83(3): 481-485.

Sinag, A., Kruse A. & Rathert J. (2004). Influence of the heating rate and the type of catalyst on the formation of key intermediates and on the generation of gases during hydropyrolysis of glucose in supercritical water in a batch reactor. *Ind. Eng. Chem. Res* 43(2): 502-508.

Sinag, A., Kruse A. & Schwarzkopf V. (2003). Key compounds of the hydropyrolysis of glucose in supercritical water in the presence of K2CO3. *Ind. Eng. Chem. Res* 42(15): 3516-3521.

Su, L., Wu X. H., Liu X. R., Chen L. Y., Chen K. Y. & Hong S. M. (2007). Effect of increasing course of temperature and pressure on polypropylene degradation in supercritical water. *Chinese Journal of Chemical Engineering* 15(5): 738-741.

Su, X. L., Zhao Y. L., Zhang R. & Bi J. C. (2004). Investigation on degradation of polyethylene to oils in supercritical water. *Fuel Processing Technology* 85(8-10): 1249-1258.

Sutton, D., Kelleher B. & Ross J. R. H. (2001). Review of literature on catalysts for biomass gasification. *Fuel Process Technol* 73(3): 155-173.

Takeshita, Y., Kato K., Takahashi K., Sato Y. & Nishi S. (2004). Basic study on treatment of waste polyvinyl chloride plastics by hydrothermal decomposition in subcritical and supercritical regions. *Journal of Supercritical Fluids* 31(2): 185-193.

Usui, Y., Minowa T., Inoue S. & Ogi T. (2000). Selective Hydrogen Production from Cellulose at Low Temperature Catalyzed by Supported Group 10 Metal. *Chemistry Letters* 29(10): 1166-1167.

Watanabe, M., Hirakoso H., Sawamoto S., Tadafumi A. & Arai K. (1998). Polyethylene conversion in supercritical water. *Journal of Supercritical Fluids, The* 13(1-3): 247-252.

Watanabe, M., Inomata H. & Arai K. (2002). Catalytic hydrogen generation from biomass (glucose and cellulose) with ZrO2 in supercritical water. *Biomass Bioenerg* 22(5): 405-410.

Watanabe, M., Inomata H., Osada M., Sato T., Adschiri T. & Arai K. (2003). Catalytic effects of NaOH and ZrO2 for partial oxidative gasification of n-hexadecane and lignin in supercritical water. *Fuel* 82(5): 545-552.

Watanabe, M., Mochiduki M., Sawamoto S., Adschiri T. & Arai K. (2001). Partial oxidation of n-hexadecane and polyethylene in supercritical water. *J Supercritical Fluids* 20(3): 257-266.

Xu, X. D. & Antal M. J. (1998). Gasification of sewage sludge and other biomass for hydrogen production in supercritical water. *Environmental Progress* 17(4): 215-220.

Xu, X. D., Matsumura Y., Stenberg J. & Antal M. J. (1996). Carbon-catalyzed gasification of organic feedstocks in supercritical water. *Ind Eng Chem Res* 35(8): 2522-2530.

Yan, Q. H., Guo L. J. & Lu Y. J. (2006). Thermodynamic analysis of hydrogen production from biomass gasification in supercritical water. *ENERGY CONVERS. MANAGE.* 47(11-12): 1515-1528.

Yoshida, T. & Matsumura Y. (2001). Gasification of cellulose, xylan, and lignin mixtures in supercritical water. *Ind. Eng. Chem. Res* 40(23): 5469-5474.

Yoshida, T., Oshima Y. & Matsumura Y. (2003). Gasification of biomass model compounds and real biomass in supercritical water. *Biomass Bioenerg* 26(1): 71-78.

# 10

# Wastewater Protozoan-Driven Environmental Processes for the Protection of Water Sources

Momba MNB
*Department of Environmental, Water and Earth Sciences,*
*Private Bag x 680 Arcadia Campus, Pretoria 0001,*
*South Africa*

## 1. Introduction

Since the middle of the last century, while the population doubled, water use has tripled. The world's freshwater reserves are dropping year after year and becoming a scarce commodity. On the one hand, the utilization of synthetic detergents, the increased use of agricultural inorganic fertilizers and the population explosion have led to the acceleration of the process of pollution of lakes and other surface water. Reports have shown an increasing percentage of rivers and streams that are contaminated with chemical and microbial pollutants (Momba et al., 2004; 2005; Bahlaoui et al., 1997; Clarke et al., 2008; Pennil et al., 2008). This is compounded by the alterations in the hydrological cycle, associated with the global climate change that increases the magnitude and frequency of runoff events (Rose et al., 2001). On the other hand, the world is faced with problems related to the management of wastewater due to the extensive industrialization, increasing population density and a highly urbanized society (McCasland et al., 2008). Recycling municipal and industrial wastewater is therefore essential for reducing the negative impact of pollution on the freshwater reserves and also for protecting public health by safeguarding water supplies against the spread of waterborne diseases (Bitton, 1999).
The first part of this chapter discusses the importance of wastewater treatment for the protection of water resources. The second part sheds light on the role protozoa play in the excess removal of phosphate and nitrate in wastewater treatment plants, with emphasis on the removal efficiency of two ciliates (*Aspidisca, Trachelophylum*) and one flagellate (*Peranema*). The third part reveals the predation potential of these protozoan species on pollution indicators and pathogenic bacteria (*Escherichia coli* O157:H7, *Salmonella typhimurium, & Shigella flexneri* spp.).

## 2. Wastewater treatment for the protection of water resources

Before wastewater treatment was required, raw wastewater was discharged directly into streams and lakes. Despite significant and prolonged environment damage, major environment legislation did not take effect until the 1970s (Gerardi & Zimmerman, 2005). To maximize the health and environmental benefits associated with the use and discharge of wastewater, great deals of legislation and several guidelines have been developed. The World Health Organization (WHO) Guidelines for the use of effluents were developed in

1973, with revised editions in 1989 and 2006. The World Health Organisation (WHO) establishes the limit for nitrates in drinking water at 50 mg NO3/L (or 11.3 mg NO3-N/L). However, the US Environmental Protection Agency (EPA) has set this limit at 10 mg NO3/L (2.3 mg NO3-N/L) (WHO, 2006).

Because guidelines must be considered in the context of national environmental, social, economic and cultural conditions, several countries have formulated their own regulations. In the United State of America, the Clean Water Act was amended in 2002. The objective of the Act is to restore and maintain the chemical, physical and biological integrity of the United States water and to prevent, reduce and eliminate pollution (Clean Water Act, 2002). The European Community, on the other hand, has established the maximum admissible levels of nitrates plus nitrites in drinking water at 50 mg NO3/L (Directive 98/83/EC). With regard to discharges, the admissible concentration depends on the receiving environment, and usually ranges between 10 and 30 mg NO3-N/L for discharges into fresh water and 50 mg NO3-N/L for discharges into seawater. These limits are lower if the discharges occur in "sensitive areas", ranging between 10 and 15 mg Total-N/L (Directive 91/271/EEC). In South Africa, the National Water Act of 1998 was enacted. The Act makes provision for the legal requirements, registration and authorization for the discharge of wastewater into a water sources (National Water Act, 1998). Also in the Act are the general wastewater discharge limits. According to the Act, there must be compliance with the set limits before wastewater can be discharged into receiving water bodies or used for irrigation. The Act of 1998 established the limits ranging between 2 and 6 mg/L for ammonia nitrogen, between 1.5 mg/L and 15 mg/L for nitrate/nitrite and between 1 and 10 mg/L for orthophosphate as phosphorus in the effluent discharge. These guidelines are an integrated and preventive management framework for maximizing the public health benefits of wastewater use and discharge.

Wastewater collection and treatment systems have been designed and built for the following purposes: (i) to provide clean water for cities undergoing rapid industrialization, (ii) to protect the quality of the waters receiving the effluent from wastewater treatment plants, and (iii) to control the outbreaks of communicable diseases. Outbreaks of communicable diseases are often related to poor sanitary conditions (Gerardi and Zimmerman, 2005). This section focuses on the characteristics of wastewater and the most common biological processes used for wastewater treatment, with the emphasis on nitrogen and phosphorus removal.

### 2.1 Characteristics of wastewater

The chemical characteristics of wastewater that are of special concern include pH, dissolved oxygen (DO), oxygen demand (chemical and biological), nitrogen (nitrite and ammonia), phosphate and metals (heavy and trace) (Larsdotter, 2006). The pH values of lower than 5 and greater than 10 indicate the presence of industrial waste and non-compatibility with biological operations (Gray, 2002). The DO is required for the respiration of aerobic microorganisms as well as other aerobic life forms. Eutrophication due to excessive amounts of nutrients contributes to the depletion of dissolved oxygen (Momba et al., 1997). The high concentration of inorganic phosphate and nitrogen in lakes and other aquatic environments is often the major cause of algal and bacterial blooms. These degrade the quality of water sources by giving the water an offensive appearance, odour and taste, and they consume oxygen. The discharge of nitrogen into receiving waters exerts a high oxygen demand at night, which adversely affects fish and other aquatic life, and has a negative impact on the beneficial use of water resources for drinking or recreation. Nitrogen compounds discharged into the environment can also cause hazards to human health. Furthermore, nitrates can form nitrosamines and nitrosamides, potentially carcinogenic compounds (Ono et al., 2000). The prevention of eutrophication can be achieved by removing phosphate and nitrogen from wastewater by chemical and biological methods or a combination of the two. The chemical removal used in most countries is expensive and results in the accumulation of large quantities of chemical waste sludge. This disadvantage has led to intensified investigations into biological phosphate and nitrogen removals, which provide a more cost-effective alternative to chemical treatment methods.

Wastewater not only constitutes an important source of chemical pollutants, but also a reservoir of pathogenic microorganisms. The major microorganisms found in wastewater influents are viruses, bacteria, fungi, protozoa and helminthes. On a daily basis, a large number and diversity of pathogenic disease-causing organisms enter sanitary sewer systems and wastewater treatment plants. These pathogens enter sewer systems from domestic wastewater and industrial wastewaters through inflow and infiltration. Pathogenic bacteria, viruses and protozoan parasites can be found in the sewer system in the wastewater, sediment, and biofilm and at wastewater treatment plants in wastewater, sludges, bioaerosols, contaminated surfaces, foam, recycle streams and scum (Gerardi, 2006, Momba & Sibewu, 2009). The biological removal mechanisms of pathogenic organisms include antibiosis, exposure to biocides, predations, and attack by litic bacteria, natural die-off and competition for limiting nutrients or trace elements (Green et al., 1997). Predation of bacteria by protozoa has been reported to be the main mechanism that contribute to the removal of bacteria in wastewater treatment plants, and, more importantly, ciliates have been actually found to play a dominant role (Curds, 1992; Pauli et al., 2001).

### 2.2 Biological phosphorus and nitrogen treatment processes

The biological phosphorus and nitrogen removal systems have been extensively investigated for municipal wastewater treatment and significant advances have been made in the area of engineering (design) and technology (implementation and operation) of the single activated sludge system. A large number of biological processes have been developed for the combined removal of nitrogen and phosphorus. Many of these processes use a form of activated sludge system process, but employ a combination of anaerobic, anoxic and aerobic zones to accomplish nutrient removals. Activated full-scale sludge systems have been successfully designed and implemented to progressively include aerobic COD removal and nitrification, anoxic denitrification and anaerobic/anoxic/aerobic excess phosphorus removal. This implementation has been aided by the development of a suite of steady state design and kinetic simulation models (Ekama et al., 84; Wentzel et al., 1992).

The activated sludge process was originally developed for carbon, nitrogen and phosphorus removals. The basic process for the simultaneous biological removal of phosphorus and nitrogen was proposed by Barnard in 1976 and is known as the Phoredox process in South Africa and Bardenpho or Modified Bardenpho in the United States. The Bardenpho activated sludge process is a four-stage process designed mainly for removing nitrogen, whereas the Phoredox activated sludge process is a five-stage process designed to remove nitrogen and phosphate. Nevertheless, the two systems consist of a sequence of primary anoxic, primary aerated, secondary anoxic, and a secondary aerated basin, followed by a clarifier. In the Phoredox activated sludge process, the incorporation of an anaerobic zone at some point in the process allows the release of phosphate; and phosphate, after being released from the biomass in the anaerobic zone, is reincorporated in the biomass during aerobiosis, together part or all the influent phosphate (Gerber et al., 1986; Momba and Cloete, 1996a,b).

In addition to the Phoredox and Bardenpho processes, other biological nutrient-removal (BNR) systems have been developed and implemented. These new BNR systems include the $A^2$ /O process, the University of Cape Town or UCT process, the modified UCT process and the Virginia Initiative Process or VIP. The $A^2$ /O™ process was developed in the United States of America. It is one of the simplest biological phosphorus removal systems. The returned activated sludge is mixed with the incoming wastewater and this mixed liquor passes through an anaerobic zone and then through an aerobic zone (Muyima et al., 1995). This process provides relatively low phosphorus removal due to high nitrate recycle to the anaerobic zone. Its layout is similar to the modified Phoredox in general, but the reactor has all its stages divided into complete mixed cells. The process is used for high-loaded activated sludge systems (Rybicki, 1997). The UCT process is derived from the Phoredox process. In this process, both the return activated sludge and the aeration tank contents are recycled to the anoxic zone, and the contents of the anoxic zone are then recycled to the anaerobic zone. This recycle sequence decreases the chance of introducing residual nitrate into the anaerobic zone. The internal recycle can be controlled to maintain zero nitrates in the effluent from the anoxic reactor, thereby ensuring that no nitrates will be returned to the anaerobic reactor. For weak wastewater, the UCT process can achieve both phosphorus and partial nitrogen removal to 6.8 mg/L (Water Environment Federation, 1996; Gray, 2002). In the modified UCT process, the anoxic zone is subdivided into parts: the first anoxic zone receives sludge recycle while an anoxic-anaerobic mixed liquor recycle is taken from it. The second anoxic zone part receives aerobic-anoxic mixed liquor recycle. The advantage of this process is that the first anoxic part is designed to reduce only the nitrate nitrogen in the return sludge, which prevents nitrogen intrusion in the anaerobic zone (Muyima et al., 1995; Rybicki, 1997; Tchobanoglous, 2003). The VIP was developed by Hampton Roads Sanitation District and the engineering firm CH2M Hill. In this process, which is similar to the UCT process, all the zones are staged to consist of at least two completely mixed cells in series (Metcalf and Eddy, 2003). The process includes the anaerobic, anoxic and aerobic zones.

### 2.3 Microorganisms involved in the biological nutrient removal

The concentration of total phosphate should not exceed 2 mg/L for installations of 10,000 - 100,000 i.e., and 1 mg/L for installations designed for more than 100,000 i.e. The respective criteria for nitrogen are 15 and 10 mg/l. Member states are expected to comply for discharges above 15,000 i.e. (Bowker & Stensel, 1990). Biological wastewater treatment systems rely on the interaction and metabolisms of microorganisms. These systems essentially depend on the capacity of the microbial community to recycle elements by way of biogeochemical cycles.

The composition of the microbial community is determined by the type of nutrient content of the wastewater and a number of other selective pressures such as mean cell retention time, aerobiosis, anaerobiosis, temperature and other extrinsic factors (Cloete & Muyima, 1997). In biological wastewater treatment systems, bacteria and other microorganisms breakdown and metabolize the COD to 10-100mg/L (Eckenfelder et al., 1992). Water is aerated and microorganisms convert the organic carbon to carbon dioxide and into cell biomass. Biomass is separated from the treated wastewater in the clarifier for recycling or wasting to solids-handling process.

Over the past three decades, most investigations have been concentrated on the important role of bacteria in the biological treatment of wastewater influents. The bacterial population has become important in the operation and control of the biological phosphorus and nitrogen removal processes. Parallel to the development in engineering and technology of the activated sludge system, significant advancements have been made in the bacteriological and biochemical analytical methods. In these areas, researchers have increasingly moved away from pure culture work to a number of new analytical techniques for the study of in situ bacterial population, for example, ATP analysis (Nelson & Lawrence, 1980), DNA analysis (Liebeskind & Dohmann, 1994), quinine profiling (Hu et al., 1999) and the use of fluorescent probes for ribosomal RNA (Wagner et al., 1994). The 16S RNA-based clones libraries or denaturing gradient gel electrophoresis have resulted in a number of high-diversity groups of bacteria involved in enhanced biological nutrient removal (Zeng et al., 2003).

Phosphate release in the anaerobic zone followed by excess phosphate uptake in the aerobic zone constitutes the main characteristics of an activated sludge system. The active biomass is returned to the reactors after settling out in a clarifier. Polyphosphate and insolubilised mineral phosphate are important fractions of activated sludge, because phosphate removal efficiency in the activated sludge process depends mainly on the sludge phosphate content. The polyphosphate is accumulated in the sludge, accompanied by potassium and magnesium accumulation. Magnesium and potassium are therefore known to be essential to enhanced phosphate removal (Toerien et al., 1990; Momba % Cloete, 1996). Sludge age is also an important factor in the removal of phosphate, because the time that the microorganisms are given to break down the waste products has a significant effect on effluent quality. Sufficient time must be permitted for microorganisms to be in contact with the waste to accomplish the treatment. A sludge age ranging from 15 to 20 days is therefore required for the removal of phosphate in activated sludge systems (Toerien, 1990). Previous investigators have also shown the role of carbon sources in nutrient removal from wastewater (Kargi et al., 2005; Akpor et al., 2008).

Studies have shown that *Acinetobacter* sp. is of little significance in phosphate removal when compared to members of other phylogenic groups of bacteria (Momba & Cloete, 1996a,b; Band et al,. 1999; Jeon et al., 2003). Other bacteria such as *Aeromonas, Vibrio, Pseudomonas,* and coliforms have been implicated in dominant polyphosphate-accumulating organisms (PAO) (Momba and Cloete, 1996a,b; Snaidr et al., 1997; Seviour et al; 2003). A series of studies by Momba and Cloete (1996a,b) on the relationship between biomass concentrations and phosphate uptake have demonstrated the role of initial biomass concentration of PAO bacteria such *Acinetobacter junii, A. radioresistens, Pseudomonas fluorescens* and *Escherichia coli* to remove phosphate from a mixed liquor medium in a laboratory-activated sludge scale system using different initial biomass cell concentrations (from $10^4$ to $10^8$ cells/mL). In this study, phosphate removal was biomass and growth-stage related. The results showed a relationship between a high initial cell density and phosphate uptake. *Acinetobacter junii* and *P. fluorescens* at a high initial biomass concentration of $10^8$cell/mL removed all the 28.25 mg/L phosphate during the entire duration of the 24 h growth study. Low initial biomass concentrations triggered the release of phosphate once transferred into the mixed liquor. The release of phosphate increased during active growth and uptake occurred when cells reached the stationary growth phase. The rate of the phosphate removal increased during the stationary growth phase for the *A. junii* and during the logarithmic growth phase for *P. fluorescens*. Enhanced phosphate uptake in both cases was related to the final cell yield in the culture media. *Acinetobacter radioresistens*, and *E. coli* at $10^6$ cells/mL and/or $10^7$cells/mL (initial cell density) removed phosphate during the first hour of the lag growth phase (17.14mg/L and 15.14mg/L for *A. radioresistens*, 6.64mg/L for E. coli). Some accumulated phosphate was released back into the medium during active growth. Both these species removed some phosphate during the stationary growth phase. *Pseudomonas fluorescens* removed more phosphate compared to *A. radioresistens* and *E. coli* with a specific rate of 3.00 - 28.50 for *P. fluorescens* , 4.92 - 17.14 mg/L for *A. radioresistens* and 0.50 -8.50 mg/L for E. coli. Finally, the most favorable net phosphate removal from mixed liquor was obtained in all cases by using a high initial biomass concentration (Momba & Cloete, 1996a, b).

The biological nitrogen removal process generally results from the combined nitrification and denitrification processes (Wentzel, 1991; Carrera et al., 2003; Oguz, 2005). In the nitrification process, the first step is the conversion of ammonia to nitrite by Nitrosomonas, while the second step is the further oxidation of nitrite to nitrate, which is commonly accepted to be carried out by Nitrobacter (Antoniou et al., 1990; Sedlak, 1991). It has been estimated that 80 % of the energy generated by nitrifier autotrophs is used to fix carbon (iv) oxide (Painter, 1970; Prosser, 1989; Sabalowsky, 1999). Unlike autotrophic nitrification where nitrification is required in order to generate energy necessary for growth, it is generally accepted that heterotrophic nitrification is not linked to cellular growth (Prosser, 1989; Pennington & Ellis, 1993).

The biological denitrification process enables the transformation of oxidized compounds by a wide spectrum of heterotrophic bacteria that convert nitrate to harmless nitrogen gas (Foglar et al., 2005). The necessary condition for denitrification to take place in activated sludge systems is the presence of a facultative microbial mass. Many common denitrifiers found in activated sludge systems appear to be capable of heterotrophic nitrification, which appears to occur simultaneously with denitrification (Prosser, 1989; Pel et al., 1997). Common denitrifiers reported in activated sludge systems include *Achromobacter, Aerobacter, Alcaligenes, Bacillus, Brevibacterium, Denitrobacillus, Flavobacterium, Lactobacillus, Micrococcus, Brevibacterium, Pseudomonas, Spirillum, Proteus, Xanthomonas, Staphylococcus* and *Paracoccus* (Gray, 1990; Metcalf & Eddy, 1999; Sabalowsky, 1999).

The efficiency of wastewater treatment plants by activated sludge system is not only linked to the bacterial population but also to the protozoa (Nicolau et al., 1997). According to their locomotive, protozoa are commonly placed in five groups, which include amoebae, flagellates, free-swimming ciliates, crawling ciliates and stalked ciliates. Most of the protozoa found in the sludge are ciliates and they can be classified in four groups: free-swimming, crawling, attached and carnivorous. Ciliated protozoa are currently used as biotic indicators due to the fact they reduce the concentration of bacteria and suspended particles in the treatment process, resulting in the production of effluents of good qualities (Curds & Cockburn, 1970a; b; Madoni et al., 1993, Salvadö et al., 1995).

The role of protozoa in the decomposition of sewage has been reported since 1964. Johannes (1964, 1965) was the first investigator to emphasize the role of protozoa in the regeneration of phosphorus, a role traditionally assigned mainly to bacteria (Fenchel, 1986, 1988). High concentrations of ciliates and other protozoa have been found to be characteristics of decomposing sewage (Johannes, 1964; 1965; Caron et al., 1985; Andersen et al., 1986). In aquatic ecosystems, ciliates and phagotrophic microflagellates have been reported to play an important role in the regeneration and mineralization of nitrogen in large quantities while grazing (Gast & Horstmann, 1983; Andersen et al., 1986). Studies have also revealed that flagellated and ciliated protozoa account for a major portion of nitrogen recycling (uptake and excretion) in both marine and freshwater habitats (Caron & Goldman, 1988; Pace & Funke, 1991).

In normal conditions, the concentrations of protozoa are larger than $10^6$protozoa/L, and the concentration of $10^7$protozoa/L corresponds to very good pollution abatement. On the contrary, concentrations lower than $10^5$ protozoa /L are indicative of a poor efficiency of the plant (Drakides, 1978). In terms of biomass, protozoa represent between 0.17% and 0.44% of the sludge during the colonization phase but can represent up to 9% at steady-state (Madoni, 1994a). Curds and Cockburn (1970b) have established relationships between the abundance of some species and the sludge loading: they have associated them with the quality of the effluent, depending on the biological oxygen demand (BOD). A recent study by Sibewu and coworkers (2008) indicated a high diversity of protozoan population in four South African activated sludge systems, which included 68 protozoan genera made up of 44 ciliates, 16 flagellates and eight others. Although the average density of ciliates was $10^4$ cells/mL in all aerobic zones, the plants that had total protozoan genera of 27 or 26 and a $BOD_5 < 25$ mg/mL produced highly-purified effluents.

Despite the evidence that the protozoan community plays a role in the recycling of wastewater and other aquatic ecosystems, increasing knowledge about the relationship between the protozoan biomass and enhanced phosphorus and nitrogen uptake is highly important for the production of effluent of a high standard and for the protection of water sources.

## 3. Role of protozoa in the excess removal of phosphate and nitrogen in wastewater systems

The average concentration of total phosphorus (inorganic and organic forms) in wastewater is in the range of 10–20 mg/L, much of which comes from phosphate builders in detergents. Common forms of phosphorus in wastewater are orthophosphate (PO4) (50–70 percent of phosphorus), polyphosphates, and phosphorus tied to organic compounds. Orthophosphate comprises approximately 90% of phosphorus in biologically treated effluents (Meganck & Faup, 1988). The total nitrogen is comprised of organic nitrogen, ammonia nitrogen, nitrite and nitrate. The concentrations of each of the types of nitrogen in the influent wastewater treatment of wastewater treatment depend on the composition of the discharges to the collection system and the conditions in the collection system prior to entering the wastewater treatment plant. Typical range values for typical strength domestic wastewater as determined by Metcalf and Eddy (1991) are as follows: 12–50 mg/L for ammonia nitrogen, 8–35 mg/L for organic nitrogen and 20–85 mg/L for total nitrogen. Since phosphorus and nitrogen are limiting nutrients and are mainly responsible for eutrophication of surface waters, they must be removed by wastewater treatment processes before discharge of the effluents into surface waters.

**Case study- Protozoan biomass relationship to enhanced phosphate and nitrate removal**

In a recent study conducted by Akpor and Momba-workers (2010), the relationship between protozoan biomass concentrations and phosphate and nitrate removal was investigated in sterile mixed liquors obtained from the anaerobic zone of an activated sludge system. The study was carried out in a shaking flask environment using three initial biomass concentrations (10, 100 and 1000 cells/mL) of two ciliates (*Aspidisca, Trachelophyllum*) and one flagellate (*Peranema*). The three organisms were isolated from the aerobic zone of the same activated sludge system. To enhance nutrient removal from mixed liquor in the presence of the protozoan isolates, sodium acetate (5 g/L) as a carbon source and potassium (0.18 g/L KNO3.) were added to the medium. Samples were taken every 24 h to determine phosphate and nitrate for a period of 96 h. Figure 1 shows an example of the pattern of the relationship between protozoan biomass and enhanced nutrient removal.

The results of the study revealed no remarkable removal of phosphate or nitrate during the first 24 h, as there was no specific increase in protozoan densities. A drastic decrease in phosphate and nitrate contents occurred with a progressive increase in protozoan biomasses (Fig. 1). This trend was observed at all initial biomass concentrations for all protozoan species (results not shown). Between 24 and 96 h, the increases in the protozoan densities corresponded to a phosphate decrease from initial ranges of 55.42–57.36 mg/L, 50.27–51.17 mg/L and 50.01–50.83 mg/L to final ranges of 2.46–11.90 mg/L, 0.61–11.80 mg/L and 1.29–13.89 mg/L, in the presence of *Aspidisca, Trachelophyllum* and *Peranema*, respectively. Nitrate concentrations were observed to decrease from initial ranges of 23.84 to 25.90 mg/L, 23.94 to 25.84 mg/L and 26.12 to 26.54 mg/L to final ranges of 0.11 to 6.32 mg/L, 0.16 to 5.60 mg/L and 0.24 to 9.04 mg/L, in the presence of *Aspidisca, Trachelophyllum* and *Peranema*, respectively.

The COD and the DO concentrations in the mixed liquor were also taken into consideration. The study revealed no COD removal in the presence of acetate as a carbon source. At initial biomass concentration of $10^3$ cells/mL, COD increases of 50.58 % for *Aspidisca*, 53.79 % for *Trachelophyllum* and 60.18% for *Peranema* were recorded (Fig 1). However, a DO decrease of over 93 % occurred in mixed liquor from the first 24 h of the experimental study and this was irrespective of the protozoan isolates and the initial biomass concentrations (Akpor & Momba, 2010).This study by Akpor & Momba (2010) revealed a direct relationship between decreases in phosphate and nitrate concentrations and the protozoan biomass. An increase in protozoan growth automatically triggered a corresponding nutrient removal and a decrease in DO concentration. It is also important to note that the growth rates of the three protozoa in mixed liquor were dependent on the initial biomass concentration of the inoculums. This study shows the need to create an environment for the proliferation of the test protozoa in activated sludge systems (Akpor & Momba, 2010).

A progressive increase in the population of protozoa with time has been reported by Petropoulos and coworkers (2005). As stated above, Momba & Cloete (1996a; b) have also reported the relationship between bacterial biomass and enhanced phosphorus removal. These studies and the current study by Akpor & Momba (2010) demonstrate the important of high concentrations of biomass for the most favourable net nutrient removal from wastewater. Hence, the introduction of high initial biomass of bacteria and protozoan species into the activated sludge system may be one of the possible means of enhancing the phosphate and nitrate uptake rate. Therefore, it would be necessary to determine in the activated sludge system which strains of these organisms are capable of nutrient removal and which initial biomass concentrations are able to result in excess uptake of phosphate and nitrate.

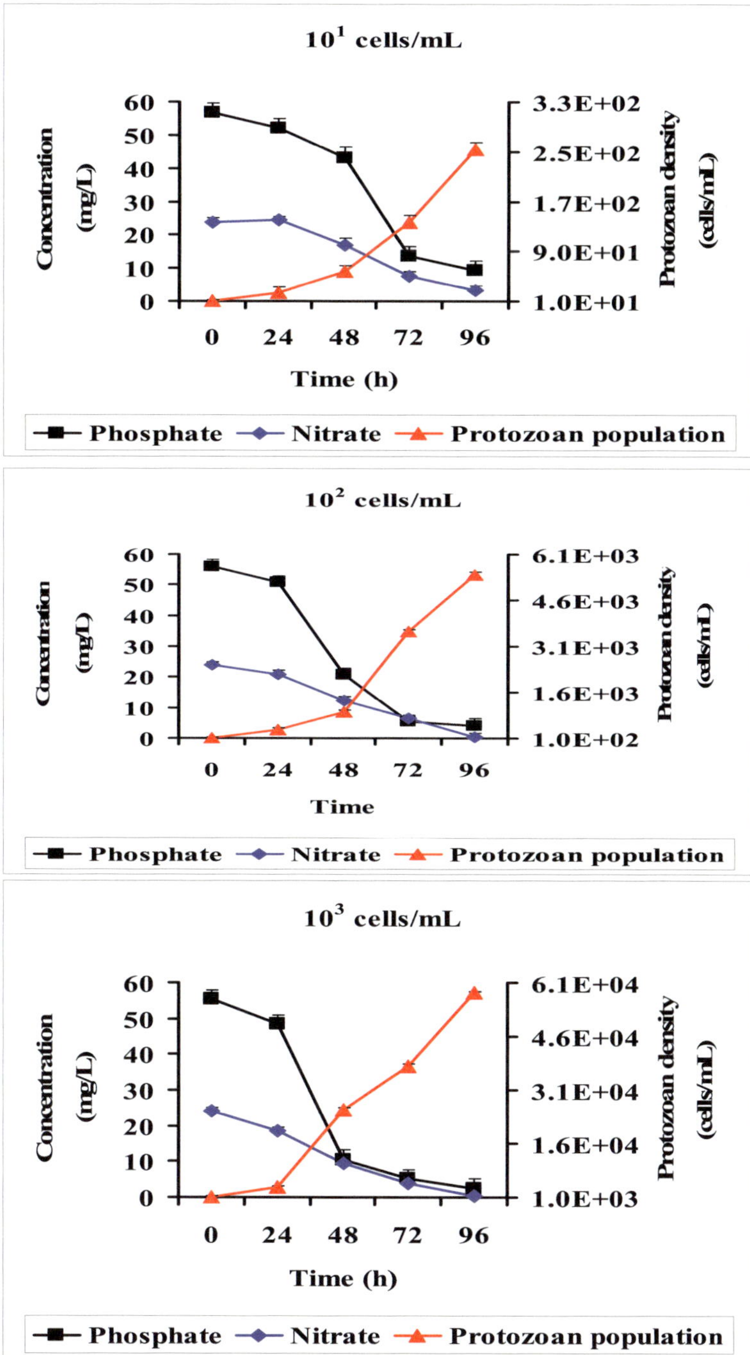


Fig. 1. An example of the relationship between nutrient (phosphate and nitrate) removal and the growth of *Aspidisca* sp in the presence of acetate as carbon source at the different initial biomass concentrations. Similar patterns were also noted for *Trachelophyllum* and*Peranema*. (Full results for these two protozoan species can be found in Akpor & Momba, 2010)

## 4. Predation potential of wastewater protozoa on human pathogenic bacteria

Previous studies have reported that the presence of protozoa in aeration tanks of wastewater treatment plants reflects an improvement in the effluent quality (Curds & Cockburn, 1970b; Sibewu et al., 2008). Researchers have focused on the significance of protozoa as reducers of bacterial concentrations and suspended particles in biological wastewater treatment plants (Curds & Cockburn, 1970a, b; Al-Shahwani & Horan, 1991; Madoni et al., 1993; Salvado et al., 1995). Two mechanisms have been proposed by which protozoan predation participates in the removal of indicator and pathogenic bacteria in biological wastewater treatment plants. The most commonly proposed mechanism is that predators graze on free bacteria and on bacteria attached to the flocks (Heinbokel, J.F., (1978). The second proposed mechanism is that suspension-feeding predators remove suspended particles as the particles flow through the wastewater treatment plant. Predators that graze on attached bacteria potentially free up the sites for future bacterial attachment, while the suspension-feeding predators directly remove particles from the mobile phase (Weber-Shirk & Dick, 1999). The feeding of flagellates and ciliates on bacteria represents the oldest predator–prey interaction that is known in nature. Flagellates and ciliates can control bacterial density in many ecosystems (Berninger et al., 1991). The majorities of investigations on the predation of bacteria by protozoa have focused either on laboratory strains of protozoa or on marine isolates (Omori and Ikeda, 1984; Laybourn-Parry et al., 2000; Scott et al., 2001; Pedros-Alio et al., 2000).

During the past decade, studies have demonstrated that different bacterial strains are not equally vulnerable to grazers and have evolved different mechanisms to resist capture, ingestion or digestion by bacterivories (Hahn & Höfle, 2001; Jürgens & Matz. 2002). This has first been documented for Legionella pneumophila, which can multiply inside amoeba such as Acanthamoeba (Omar et al., 2000). Phenotypic bacterial properties that have been identified to influence grazing mortality are size and morphology (Gonzalez et al., 1990; Šimek & Chrzanowski, 1992, Šimek et al., 2001),, swimming speed (Matz & Kjelleberg 2005), toxic pigments (Matz et al., 2004), the physico-chemical surface structure (Monger et al., 1990; Matz & Jürgens 2001) and cell-to-cell communication (Okada et al., 2005). These findings have indicated that predation by protozoa is an important factor that contributes to the diversification of bacterial traits and might select specific antipredator adaptations in bacteria. It is therefore interesting to describe this phenomenon during the biological treatment of wastewater by activated sludge. Hence the influence of protozoan grazing (grazing rate, clearance rate and ingestion rate) on pollution indicator and enteric bacteria pathogens from mixed liquor wastewater was investigated and the optimum density of protozoa required for the removal of pathogenic bacteria was identified in our laboratory. The results of this study are discussed below.

**Case study- Protozoan biomass relationship to pathogenic bacteria removal from wastewater**

To investigate the protozoan biomass relationship to pathogenic bacteria removal, three strains of human enteric pathogenic bacteria frequently found in wastewater treatment plants [(*pathogenic Salmonella typhimurium* (ATCC 14028), *Shigella flexneri* (CCRC 1077) and *E. coli* O157:H7 (ATCC 43895)] and three protozoan isolates (*Aspidisca sp., Trachelophylum sp.* and *Peranema* sp.), which were previously screened for phosphate and nitrate removal efficiency (Akpor et al., 2007) were used. For each selected protozoa, three different initial biomass concentrations ($10^2$cell/ml, $10^3$cell/ml and $10^4$cells/ml) were considered. The grazing experiments were carried out using a modification of the protocol outlined by Sherr and Sherr (1993). Each experiment consisted of duplicate batches containing one protozoa species and one of live pathogenic bacterial strains. Experiments were conducted in Erlenmeyer flasks containing 250 mL of modified wastewater mixed liquor, which is a simulated environment for the bacteria and protozoa isolated from the wastewater system (Momba & Cloete, 1996). For each protozoan species, flasks were separately inoculated with initial protozoa biomass ($10^2$,$10^3$ and $10^4$ cells/mL and then placed in the dark in a water bath at 25°C for 30 min to allow the protozoa to recover from handling shock and to acclimatise. For each pathogenic bacterium, an initial concentration of $10^6$cfu/mℓ was separately added to each flask, followed by incubation in the dark with shaking (100 ± 10 rpm) at room temperature. Five milliliter aliquots were collected from each flask on an hourly basis for the first 12 h and thereafter every 18 h and 24 h. Protozoan densities were estimated by dispensing approximately 1mL of the sample on improved Neubauer hemocytometer and viewing directly under the microscope at 400 X magnifications (Axioplan Carl Zeiss GmbH equipped with phase contrast, bright field and epifluorescence, HBO 50 illuminator and a digital imaging system). Spread plate procedure and selective media (XLD agar for *Salmonella typhimurium* and *Shigella flexneri* and Fluorocult *E coli* O157:H7 agar for E coli O157:H7) were used to determine the concentration of live bacteria. The plates were incubated at 37°C for 24h. The protozoan concentrations were calculated using the following formula:

$$\text{Number of cells/mL} = (C \times V)/(A \times D \times F) \quad (1)$$

where C is the number of organisms counted, F the number of fields counted, D (mm) the depth of the counting chamber, A ($mm^2$) the area of a field and V ($mm^3$) the volume of the counting chamber (APHA 2001). Growth rates ($\mu$) ($d^{-1}$) were calculated using the following equation:

$$\mu = (\ln F_t - \ln F_0)/\Delta t \quad (2)$$

where $F_0$ and $F_t$ denote the concentrations of the protozoa at the beginning and at the end of the time interval ($\Delta t$) of exponential increase during the study period (APHA 2001).
The grazing rates (G) (per hour) were calculated by the equation:

$$G = (N_0 - N_t)/[(N_0 - N_t)/2]/\Delta t, \quad (3)$$

where $N_0$ and $N_t$ refer to the concentrations of bacteria at the beginning and the end of the time interval ($\Delta t$) of decline. The clearance rates (C) (nl /flagellate/h) were calculated by the equation:

$$C = (G \times 1{,}000{,}000)/F_m, \quad (4)$$

where $F_m$ (Protozoa/mL) is the mean concentration of the flagellate. $F_m$ was calculated for the time of an exponential increase, according to the method of Heinbokel (1978):

$$F_m = (F_t - F_0)/(\ln F_t - \ln F_0) \quad (5)$$

where $F_0$ and $F_t$ refer to the concentrations of the protozoa at the beginning and at the end of the time interval, respectively.

The ingestion rates (I) (bacterial cells/protozoa/hour) were estimated by the equation (Wu et al., 2004):

$$I = (N_0 - N_t)/F_m/\Delta. \quad (6)$$

The results indicated a relationship between the protozoan biomass and their grazing, clearance and ingestion rates on pathogenic bacteria. During the 24 h of the predation experiments, there were no significant changes in protozoan densities, which ranged from $2 \times 10^4$ to $9.8 \times 10^4$ cells/mL, from $2 \times 10^4$ to $9.1 \times 10^4$ cells/mL and from $2 \times 10^4$ to $8.7 \times 10^4$ cells/mL when the three protozoa had *E. coli, Salmonella typhimurium* and *Shigella flexneri* as their sole sources of food, respectively. No significant differences in protozoan growth rates were noted between the initial protozoan biomass of $10^2$cells/mL and that of $10^3$cells/mL in the presence of the three enteropathogenic bacterial strains. However, significant differences in growth rates were recorded between the initial protozoan biomass of $10^4$cells/mL and those of the two former initial biomasses [P = 0.054 9 one-way ANOVA)]. For all initial protozoan densities, the maximum protozoan growth rates were observed when grazing on *E. coli*, with *Peranema* having the highest rates, ranging between 0.14 and 0.17$h^{-1}$ during the study period (Tables 1–3). The growth experiment therefore indicated a strong relationship between a higher protozoan biomass and a higher growth rate, although in various densities all the three protozoa species were able to predate successfully on the enteropathogenic bacteria, and that bacterial food concentrations of $10^6$bacteria/mL allowed a positive growth (Tables 1-3).

For all protozoan species, lower protozoan biomass concentrations of $10^2$cells/mL and $10^3$cells/mL significantly resulted in lower grazing rates (P = 0.002 [t test]), clearance rates (P = 0.002 [t test]) and ingestion rates (P = 0.007 [t test] compared to the higher protozoan biomass of $10^4$cells/mL (Tables 1–3). This clearly showed that the three mechanisms of protozoan predation on pathogenic bacteria increased with a higher protozoan biomass

| **Strain** | **Mechanisms** | *A* | *B* | *C* |
|---|---|---|---|---|
| *E. coli* O157:H7 | Grazing rate ($cell^{-1}$ $h^{-1}$) | 32 | 42 | 28 |
| | Clearance rate (nl $cell^{-1}$ $h^{-1}$) | 20 | 60.36 | 13.27 |
| | Ingestion rate (bacteria $cell^{-1}$ $h^{-1}$) | 240 | 715 | 181 |
| | Growth rate ($d^{-1}$) | 0.13± 0.01 | 0.12± 0.03 | 0.14± 0.01 |
| *Salmonella typhimurium* | Grazing rate ($cell^{-1}$ $h^{-1}$) | 4.7 | 21 | 4 |
| | Clearance rate (nl $cell^{-1}$ $h^{-1}$) | 12.89 | 52 | 12.04 |
| | Ingestion rate (bacteria $cell^{-1}$ $h^{-1}$) | 153 | 394 | 133 |
| | Growth rate ($d^{-1}$) | 0.09± 0.02 | 0.09± 0.02 | 0.06± 0.2 |
| *Shigella flexneri* | Grazing rate ($cell^{-1}$ $h^{-1}$) | 38 | 38 | 23 |
| | Clearance rate (nl $cell^{-1}$ $h^{-1}$) | 38 | 68.2 | 9.4 |
| | Ingestion rate (bacteria $cell^{-1}$ $h^{-1}$) | 366 | 502 | 148 |
| | Growth rate ($d^{-1}$) | 0.10± 0.01 | 0.10± 0.01 | 0.08± 0.02 |

*A- Aspidisca sp, B - Trachelophyllum sp and C - Peranema sp.*

Table 1. Predation of initial protozoan density of $10^2$cells/ml on enteric pathogenic bacteria

| **Strain** | **Mechanisms** | *A* | *B* | *C* |
|---|---|---|---|---|
| *E. coli* O157:H7 | Grazing rate ($cell^{-1}$ $h^{-1}$) | 36 | 51 | 33 |
| | Clearance rate (nl $cell^{-1}$ $h^{-1}$) | 21.34 | 59.4 | 15.43 |
| | Ingestion rate (bacteria $cell^{-1}$ $h^{-1}$) | 232 | 728 | 181 |
| | Growth rate ($d^{-1}$) | 0.14± 0.01 | 0.13± 0.03 | 0.15± 0.01 |
| *Salmonella typhimurium* | Grazing rate ($cell^{-1}$ $h^{-1}$) | 5 | 30 | 8 |
| | Clearance rate (nl $cell^{-1}$ $h^{-1}$) | 15 | 36 | 10 |
| | Ingestion rate (bacteria $cell^{-1}$ $h^{-1}$) | 160 | 403 | 123 |
| | Growth rate ($d^{-1}$) | 0.10± 0.02 | 0.10± 0.02 | 0.06± 0.2 |
| *Shigella flexneri* | Grazing rate ($cell^{-1}$ $h^{-1}$) | 33 | 44 | 30 |
| | Clearance rate (nl $cell^{-1}$ $h^{-1}$) | 35 | 40 | 8.9 |
| | Ingestion rate (bacteria $cell^{-1}$ $h^{-1}$) | 358 | 489 | 111 |
| | Growth rate ($d^{-1}$) | 0.11 ± 0.01 | 0.11± 0.01 | 0.09± 0.02 |

*A- Aspidisca sp, B - Trachelophyllum sp and C - Peranema sp.*

Table 2. Predation of initial protozoan density of $10^3$cells/ml on enteric pathogenic bacteria

| Strain | **Mechanism** | *A* | *B* | *C* |
|---|---|---|---|---|
| *E. coli* O157:H7 | Grazing rate ($cell^{-1}$ $h^{-1}$) | 78 | 83 | 60 |
| | Clearance rate (nl $cell^{-1}$ $h^{-1}$) | 44 | 94.1 | 60 |
| | Ingestion rate (bacteria $cell^{-1}$ $h^{-1}$) | 930 | 1035 | 440 |
| | Growth rate ($d^{-1}$) | 0.16± 0.01 | 0.16± 0.03 | 0.17± 0.01 |
| *Salmonella typhimurium* | Grazing rate ($cell^{-1}$ $h^{-1}$) | 15 | 52 | 14 |
| | Clearance rate (nl $cell^{-1}$ $h^{-1}$) | 30 | 58.6 | 77 |
| | Ingestion rate (bacteria $cell^{-1}$ $h^{-1}$) | 920 | 730 | 320 |
| | Growth rate ($d^{-1}$) | 0.11± 0.02 | 0.16± 0.02 | 0.09± 0.2 |
| *Shigella flexneri* | Grazing rate ($cell^{-1}$ $h^{-1}$) | 77 | 77 | 62 |
| | Clearance rate (nl $cell^{-1}$ $h^{-1}$) | 29 | 98.21 | 55 |
| | Ingestion rate (bacteria $cell^{-1}$ $h^{-1}$) | 540 | 1078 | 430 |
| | Growth rate ($d^{-1}$) | 0.11± 0.01 | 0.11± 0.01 | 0.15± 0.02 |

*A- Aspidisca sp, B - Trachelophyllum sp and C - Peranema sp.*

Table 3. Predation of initial protozoan density of $10^4$cells/ml on enteric pathogenic bacteria

concentration. The results of this study showed that the ingestion rate was significantly higher ($P< 0.001$) compared to the grazing rate and to the clearance rate. This was irrespective of the protozoan initial biomass concentrations. In general, the highest grazing, clearance and ingestion rates were significantly noted with *Trachelophylum* when predating on *E. coli* and *Shigella flexneri* compared to *Aspidisca* and *Peranema* (test, $P \leq 0.002$ in all cases) (Tables 1–3).

In terms of food preference, *Escherichia coli* O157:H7 remained the most preferred prey for both ciliates and the flagellate, followed by *Shigella flexneri*, and *Salmonella typhimurium* was found to be the least preferred prey among protozoa in most cases. This study clearly demonstrated that protozoa generally play a big role in the removal of indicators and pathogenic bacteria in activated sludge systems. To reveal whether this holds true for the majority of bacteria, more investigations are needed.

## 5. Conclusion

The various perspectives on the role of microorganisms in water and water resources have been highlighted in order to gain a deeper understanding of the intricate web of relationships between man, microbes and the environment. Microorganisms do not exist in a vacuum. They interact with the environment, causing harmful and beneficial effects which are in a delicate balance.

It is well known that wastewater treatment consists of physical, chemical and biological treatment processes. All these processes may be required to produce effluent of a high quality that can result in the protection of water sources. Individual wastewater treatment processes cannot be applied in a universal fashion. Each process has limitations and should be used in combination with several other processes to deliver effluent of an appropriate quality.

## 6. References

Akpor, O.B.; Momba, M.N.B. & Okonkwo, J. (2008). Effect of nutrient/carbon supplement on biological phosphate and nitrate uptake by protozoa isolates. *Journal of Applied Sciences* 8(3), 489-495.

Akpor OB and MNB Momba (2010) Protozoan biomass relationship to nutrient removal in activated sludge mixed liquor. *Biotechnology Journal.* 5, 304–313

Andersen, O.K.; Goldman, J.C.; Caron, D.A. & Dennette, M.R. 1986. Nutrient cycling in a microflagellate food chain III: phosphorus dynamics. *Marine Ecology Progress Series* 31, 47-55.

Al-Shahwani, S.M. and Horan, N.J. (1991). The use of protozoa to indicate changes in the performance of activated sludge plants. *Water Research* 25: 633–638.

Antoniou P.; Hamilton, J.; Koopman, B.; Jahi, R.; Holloway, B.; Lybertos, G. & Svoronos, S.A. 1990. Effect of temperature and pH on the effective maximum specific growth rate of nitrifying bacteria. *Water Research* 24(1), 97-101.

APHA. 2001. *Standard Methods for the Examination of Water and Wastewater*, 20th edition. APHA, Washington D.C.

Bahlaoui, M.A., Baleux, B. and Troussellier, M. (1997) Dynamic of pollution-indicator and pathogenic bacteria in high-rate oxidation wastewater treatment ponds. *Water Research,* 31(3): 574-582.

Band, P.L.; Erhart, R.; Wagner, M.; Keller, J. & Blackall, L.L. (1999). Identification of some of the major groups of bacteria in efficient and non efficient biological phosphorus removal in activated sludge systems. *Applied and Environmental Microbiology,* 65(9), 4077-4084.

Barnard, J.L. (197. Biological nutrient removal without the addition of chemicals. *Water Research* 9(5-6), 485-490.

Berninger, U.G., Finlay B.J. and Kuuppo-Leinikki P. (1991). Protozoan control of bacterial abundances in freshwater. *Limnol. Oceanogr.* 36:139–147.

Bitton, G. (1994) *Wastewater microbiology*. New York: Miley–Liss Inc.

Caron, D.A. & Goldman, J.C. (1988). Dynamics of protistan carbon and nutrient cycling. *Journal of Protozoology* 35, 247-249.

Caron, D.A.; Goldman, J.C.; Anderson, O.K. & Dennett, M.R. 1985. Nutrient cycling in a microflagellate food chain II: population dynamics and carbon cycling. *Marine Ecology Progress Series* 24, 243-254.

Carrera J.; Baeza, J.A.; Vincent, T. & Lafuente, F.J. (2003). Biological nitrogen removal of high-strength ammonium industrial wastewater with two-sludge system. *Water Research* 37, 4211-4221.

Clark ME, Giddings CW and Mcevoy JM (2008) Sources of *Cryptosporidium* in a rural watershed. International workshops on opportunistic protists. Boston, MA, May 2008.

Clean Water Act. (2002). *The Federal Water Pollution Control Act*. US Congress, USC. 1251.

Cloete T.E. & Muyima N.Y.O (1997) Microbial community analysis: The key to the design of biological wastewater treatment systems. *Scientificand Technical Report* No5. International Assoiciation on water Quality in its scientific and technical Report.Great Britain. ISBN 1 900222 027 pp 98.

Curds C.R., Cockburn A., Vandike J.M. (1968) An experimental study of the role of the ciliated protozoa in the activated-sludge process. *Water Pollution Control*. 67 , 312-329

Curds C.R.& Cockburn A .(1970a) Protozoa in biological sewage-treatment process - I. A survey of the protozoan fauna of British percolating filters and activated-sludge plants, *Water Research*, 4: 225-236.

Curds C.R. & Cockburn A (1970b). Protozoa in Biological Sewage-Treatment Process - II. Protozoa as indicator in the activated-sludge process, *Water Research* 4: 237-249.

Curds, C.R., (1992). Protozoa and the water industry. Cambridge University Press, Cambridge, UK.

Drakides C. (1978) L'Observation microscopique des boues activées appliquée à la surveillance des installations d'épurations: Technique d'étude et interprétation, *TSM-L'EAU*, 2, 85-98

Ekama, G.A.; Marias, G.R.; Siebritz, I.P.; Pitman, A.R.; Keay, G.P.; Buchan, L.; Gerber, A. & Smollen, M. (1984). Theory, design and operation of nutrient removal activated sludge processes. *University of Cape Town, City of Johannesburg and the National Institute for Water Research of the CSIR, Pretoria, South Africa.*

Eckenfelder, W.W. & Grau, P. (1992). *Activated Sludge Process Design and Control: Theory and Practice,* vol 1. Technomic \Publishing Inc, Pennsylvania, USA.

EU (1991) Council Directive Concerning Urban Wastewater Treatment. 91/271 EEC of May 21, 1991, OJ NO L135/ 40 of May 30, 1991.

Fenchel, T. 1986. The ecology of heterotrophic microflagellates. *Advances in Microbial Ecology* 9, 57-67.

Fenchel, T. 1988. Microfauna in pelagic food chains: *In Nutrient Cycling in Coastal Marine Environments.* Blackburn, T.H. & Sorensen, J. (eds). John Wiley & Sons Ltd, pp 59-64.

Gast, V. & Horstmann, U. 1983. N-remineralization of phyto and bacterioplankton by the marine ciliate *Euplotes vannus. Marine Ecology Progress Series* 13; 55-60.

Gerardi M.H & Mel C. Zimmerman M.C.(2005) *Wastewater Pathogens* John Wiley & Sons, Inc pp190 ISBN 0-471-20692-X .

Gerardi M.H (2006) *Wastewater Bacteria.* John Wiley & Sons, Inc. ISBN-13: 978-0-471-20691-0 . ISBN-10: 0-471-20691-1

Gerber A., Mostert E.S., Winter C.T. & de Villiers R.H (1986) The effect of acetate and other short-chain carbon compound on the kinetics of biological nutrient removal. *Water SA,* 12 (1): 7-12.

Gonzalez, J. M., Iriberri, J., Egea, L., and Barcina, I. (1990) Differential rates of digestion of bacteria by freshwater and marine phagotrophic protozoa. *Applied Enviromental Microbiology,* 56:1851–1857.

Gray, F.N. (2002). *Water Technology: An Introduction for Environmental Scientists and Engineers.* Butterworth-Heinemann. Oxford pp 35-80

Gray, F.N.1990. *Activated Sludge: Theory and Practice.* Oxford University Press, New York, pp 20-67.

Green, M.B., Griffin, P., Seabridge, J.K., Dhobie, D., 1997. Removal of bacteria in subsurface flow wet-lands. *Water Science &.Technology* 35 (5), 109–116.

Hahn MW & Höfle MG (2001) Grazing of protozoa and its effect on populations of aquatic bacteria. *FEMS Microbial Ecology.* 35: 113–121.

Heinbokel, J.F., (1978) Studies on the functional role of tintinnids in the Southern California Bight. I. Grazing and growth rates in laboratory cultures. Marine Biology, 47:177–189.

Hu, H.Y., Goto, N., Fujie, K., (1999). Concepts and methodologies to minimize pollutant discharge for zero-emission production. Water Science and Technology 39 (10–11), 9–16.

Jeon, C.O.; Lee, D.S. & Parker, J.M. (2003). Microbial communities in activated sludge performing enhanced biological phosphorus removal in sequencing batch reactor. *Water Research,* 37: 2115-2205.

Johannes, R.E. (1964). Phosphorus excretion and body size in marine animals: Microzooplankton and nutrient regeneration. *Science* 146: 923-924.

Johannes, R.E. (1965). Influence of marine protozoa on nutrient regeneration. *Limnology and Oceanography* 10(3), 434-442.

Jürgens, K., & Matz C. 2002. Predation as a shaping force for the phenotypic and genotypic composition of planktonic bacteria. *Antonie van Leeuwenhoek* 81:413-434.

Kargi, F.; Uygur, A. & Baskaya, H.B. (2005). Phosphate uptake and release rates with different carbon sources in biological nutrient removal using a SBR. *Journal of Environmental Management* 76, 71-75.

Larsdotter, K. (2006). Microalgae for phosphorus removal from wastewater in a Nordic climate. *A Doctoral Thesis from the School of Biotechnology, Royal Institute of Technology, Stockholm, Sweden,* ISBN: 91-7178-288-5.

Laybourn-Parry, J., Mell, E.M. and Roberts, E.C., (2000) Protozoan growth rates in Antarctic lakes. *Polar Biology,* 23: 445–451.

Liebeskind, M. & Dohmann, M. 1994. Improved method of activated sludge biomass determination. *Water Science and Technology* 29(7), 7-13.

McMahon, K.D.; Dojka, M.A.; Pace, N.R.; Jenkins, D. & Keasling, J.D. (2002). Poyphosphate kinase from activated sludge performing enhanced biological phosphorus removal. *Applied and Environmental Microbiology* 68(10), 4971-4978.

Madoni P. (1994), Estimates of Ciliated Protozoa Biomass in Activated Sludge and Biofilm, *Bioresource Technology,* 48: 254-249

Madoni P, Davoli D. & Chierici (1993). Comparative analysis of the activated sludge microfauna in several sewage treatment wor ks. Water Research, 27: 1485-1491

Matz, C. and Kjelleberg, S. (2005). Off the hook: how bacteria survive protozoan grazing. *Trends in Microbiology,* 13(7): 302-307

McCasland, M.; Trautmann, N.; Porter, K. & Wagenet, R.( 2008). Nitrate: Health effects in drinking water. Available from http://pmep.cee.comell.edu/facts.slides-self/facts/nit-heef-grw85.html. Accessed 05/04/2008.

Metcalf and Eddy (1991). *Wastewater Engineering. Treatment, Disposal, Reuse.* 3rd edition, McGraw-Hill Int. Ed., Singapore. 31pages

Metcalf, X. & Eddy, X. (1999). *Advanced Wastewater Treatment: In Wastewater Engineering, Treatment, Disposal and Re-use,* 3rd edition. McGraw-Hill New York.

Metcalf, X. & Eddy, X. (2003). *Wastewater Engineering: Treatment and Reuse. In: Wastewater Engineering, Treatment, Disposal and Re-use.* Techobanoglous, G.; Burton, F.L. & Stensel, H.D. (eds), Tata McGraw-Hill Publishing Company Limited, 4th edition. New Delhi, India.

Momba, M.N.B. & Cloete, T.E. (1996a). The relationship of biomass to phosphate uptake by *Acinetobacter junii* activated sludge mixed liquor. *Water Research* 30(2), 364-370.

Momba, M.N.B. & Cloete, T.E. (1996b). Biomass relationship to growth and phosphate uptake of *Pseudomonas fluorescens, Escherichia coli* and *Acinetobacter radioresistens* in mixed liquor medium. *Journal of Industrial Microbiology* 16, 364-369.

Momba, M.N.B. & Mfenyana, C. 2005. Inadequate treatment of wastewater: A source of coliform bacteria in receiving surface water bodies in developing countries- Case Study: Eastern Cape Province of South Africa. In Water *Encyclopedia- Domestic, Municipal and Industrial Water Supply and Waste Disposal.* Lehr, J.H. & Keeley, J. (eds). John Wiley & Sons Inc pp 661-667.

Momba MNB & M Sibewu (2009) Survival of somatic and F-RNA coliphages in treated wastewater effluents and their impact on viral quality of the receiving water bodies in the Eastern Cape Province. *Journal of Biological Sciences 9(7) 648-654*

Monger, B. C., Landry, M. R. (1990). Direct-interception feeding by marine zooflagellates: the importance of surface and hydrodynamic forces. Marine Ecology. Prog. Ser. 65: 123-140

Muyima, N.; Momba, M.N.B. & Cloete, T.E. (1995). Biological methods for the treatment of wastewaters: In Microbial Analysis: The key to the design of biological wastewater treatment systems, Cloete, T.E. & Muyima, N. (eds). *IAWA Scientific and Technical Report* No. 5.

National Water Act. (1998). *Act No 36 of 1998*. Department of Water Affairs and Forestry, South Africa.

Nicolau A., Lima N., Mota N., Madoni P., Os Protozoários como Indicadores da Qualidade das Lamas Activadas, Boletim de Biotecnologia, 56, 14-19 (1997).

Oguz, M.T. (2005). Investigation of nitrifying bacterial activities by monitoring nitrate reduction, nitrate formation and carbon dioxide fixation during activated sludge treatment in the presence of metabolic inhibitors allythiourea and azide. *Erciyes Universitesi Fen Bilimleri Enstitusu Dergisi* 21(1-2), 154-165.

Omori, M. and Ikeda, T., (1984). Methods in marine zooplankton ecology. Wiley.

Ono, Y., Somiya, I., Oda, Y., 2000. Identification of a carcinogenic heterocyclic amine in river water. Water Research, 34 (3), 890–894.

Pace, M.L.; & Funke, E. (1991). Regulation of planktonic microbial communities by nutrients and herbivores. *Ecology* 72(3),904-914.

Painter, H.A. (1970). A review of literature on inorganic nitrogen metabolism in microorganisms. *Water Research* 3, 241-250.

Pauli W., Jax, K. & Berger, S. (2001). Protozoa in wastewater treatment: function and importance. In: Beek, B. (Ed.), Biodegradation and Persistence. Springer, Berlin, pp. 202–252.

Pel, R.; Oldenhuis, R; Brand W.; Vos, A.; Gottschal, J.C. & Zwart, K.B. (1997). Stable-isotope analysis of a combined nitrification and denitrification sustained by thermophilic metanotrophs under low-oxygen condition. *Applied and EnvironmentalMicrobiology* 63(2), 472-481.

Pennington, P.I. & Ellis, R.C. 1993. Autotrophic and heterotrophic nitrification in Acidic forest and native grassland soils. *Soil Biology and Biochemistry* 25(10), 1399-1408.

Pennil CC, Clark ME, Giddings CW & Mcevoy JM (2008) Sources of *Cryptosporidium* in a rural watershed. National Water Quality Conference, Sparks, Nevada.

Pedros-Alio, C., Calderon-Paz, J.I., MacLean, M.H., Medina, G., Marrase, C., Gasol, J.M. and Guixa-Boixereu, N., (2000) The microbial food web along salinity gradients. *Microbial Ecology*. 32: 143–155.

Petropoulos, P. & Gilbride, K.A. (2005). Nitrification in activated sludge batch reactors is linked to protozoan grazing of the bacterial population. *Canadian Journal of Microbiology* 51, 791-799.

Prosser, J.I. (1989). Autotrophic nitrification in bacteria: *In Advances Microbial Ecology,* Vol. 30. Rose A.H & Tempest, D.W. (eds). Academic Press, New York pp 125-181.

Nelson, P.O. & Lawrence, A.W. 1980. Microbial viability measurements and activated sludge kinetics. *Water Research* 14, 217-225.

Rybicki, S. (1997). *Advanced Wastewater Treatment: Phosphorus Removal from Wastewater.* Royal Institute of Technology, Stockholm, Sweden Report No.1.

R.P.G Bowker and H.D. Stensel. Phosphorus removal from wastewater. Noyes Data Corporation, (1990).

Rose, JB, Epstein PR, Lipp EK and Sherman BH (2001) Climate variability and change in the United States: potential impacts on water- and foodborne diseases caused by microbiologic agents, *Environmental Health Perspectives* 109, 211-221.

Sabalowsky, A.R. (1999). An investigation of the feasibility of nitrification and denitrification of a complex industrial wastewater with high seasonal temperatures. *Master's thesis from Virginia Polytechnic Institute and State University. Blacksburg.*

Salvadó, H., Gracia, M. P. and Amigó, J. M., (1995) Capability of ciliated protozoa as indicators of effluent quality in activated sludge plants. *Water Research,* 29: 1041-1050.

Scott, F.J., Davidson A.T. and Marchant, H.J. (2001) Grazing by the Antarctic sea ice ciliate *Pseudocohnilembus. Polar Biol.* 24: 127–131.

Sedlak, R. (Ed). (1991). *Phosphorus and Nitrogen Removal from Municipal Wastewater: Principles and Practice,* 2nd edition. Lewis Publishers. New York.

Seviour, R.J.; Mino, T. & Onuki, M. (2003). The microbiology of biological phosphorus removal in activated sludge systems. *FEMS Microbiological Review* 27, 99-127.

Sherr, B.F., Sherr, E.B. and Berman, T. (1983) Grazing, growth and ammonia excretion rates of a heterotrophic microflagellate fed with four species of bacteria. *Applied and Environmental Microbiology,* 45:1196–1201.

Sibewu, M.; Momba, M.N.B. & Okoh, A.I. (2008). Protozoan fauna and abundance in Aeration tanks of three municipal wastewater treatment plants in the Eastern Cape Province of South Africa. *Journal of Applied Sciences* 8(11), 2112-2117.

Šimek K & Chrzanowski TH (1992) Direct and indirect evidence of size-selective grazing on pelagic bacteria by freshwater nanoflagellates. *Applied Environment &. Microbiology.* 58: 3715–3720.

Šimek, K., J. Pernthaler, M. G. Weinbauer, K. Hornak, J. R. Dolan, J. Nedoma, M. Masin, and R. Amann. (2001). Changes in bacterial community composition and dynamics and viral mortality rates associated with enhanced flagellate grazing in a mesoeutrophic reservoir. *Appl. Environ. Microbiol.* 67:2723-2733.

Snaidr, J.; Amann, R.; Huber, I.; Ludwig, W. & Schleifer, K.H. (1997). Phylogenetic analysis and *in situ* identification of bacteria in activated sludge. *Applied and Environmental Microbiology* 67(7), 2884-2896.

Tchobanoglous, G.; Burton, F.L. & Stensel, H.D. (2003). *Wastewater Engineering: Treatment Disposal Reuse.* Metcalf & Eddy, Inc., 4th Edition, McGrwa-Hill Books Company. ISBN 0-07-041878-0.

Toerien, D.F.; Gerber, A.; Lotter, L.H. & Cloete, T.E. (1990). Enhanced biological phosphorus removal in activated sludge. *Advances in Microbial Ecology* 11: 173-230

Wagner, M.; Erhart, R.; Manz, W.; Amann, R.; Lemmer, H.; Wedi, D.; & Schleifer, K.H. (1994). Development of a ribosomal-

RNA-targeted oligonucleotide probe specific for the genus *Acinetobacter* and its application for *in situ* monitoring in activated sludge. *Applied and Environmental Microbiology* 60(3), 2-800.

Water Environment Federation. (1996). *Operation of Municipal Wastewater Treatment Plant: Manual of Practice,* 5th edition vol. 2. Alexandria.

Weber-Shirk, M.L., Dick, R.I., 1999. Bacterivory by chrysophyte in slow sand filters. Water Research, 33 (3), 631–638.

Wentzel, M.C. 1991. Future of Nutrient Removal in South Africa. *WISA Nutrient Removal Technical Division,* 1st open meeting,

Wenzel, MC.. Ekama, GA. (1997). *Principles in the design of single sludge activated sludge systems for biological removal of carbon, nitrogen and phosphorus.* In La dephosphatation des eaux usees, Ed. CEBEDOC, Belgium, 13-26. South Africa.

WHO (2006). *Guidelines for the Safe Use of Wastewater, Excreta and Greater.* Vol. 3. World Health Organisation Press, Geneva, Switzerland.

Zeng, R.J.; Lemaire, R.; Yuan, Z. & Keller, J. (2003). Simultaneous nitrification, denitrification and phosphorus removal in a lab-scale sequencing batch reactor. *Biotechnology and Bioengineering* 84(2), 170-178.